# Ecological Studies

Analysis and Synthesis

---

Edited by

W. D. Billings, Durham (USA) F. Golley, Athens (USA)

O. L. Lange, Würzburg (FRG) J. S. Olson, Oak Ridge (USA)

Volume 20

F. B. Christiansen T. M. Fenchel

# Theories of Populations in Biological Communities

With 68 Figures

Springer-Verlag Berlin Heidelberg New York 1977

Dr. FREDDY B. CHRISTIANSEN
Professor TOM M. FENCHEL
Institute of Ecology and Genetics
University of Aarhus
Ny Munkegade, Building 550
DK–8000 Aarhus C, Denmark

For explanation of the cover motive see legend to Fig. 40 (p. 71)

ISBN 3-540-08010-4 Springer-Verlag Berlin · Heidelberg · New York
ISBN 0-387-08010-4 Springer-Verlag New York · Heidelberg · Berlin

Library of Congress Cataloging in Publication Data. Christiansen, Freddy B. 1946—. Theories of populations in biological communities. (Ecological studies; v. 20) Bibliography: p. Includes index. 1. Biotic communities. 2. Population biology. I. Fenchel, Tom, joint author. II. Title. III. Series. QH541.C39.574.5'24.76-49871.



Typesetting, printing, and binding: Brühlsche Universitätsdruckerei, Gießen. 43210

# Preface

When we wrote this book it was, admittedly, first of all for the sake of our own enjoyment and enlightenment. We will, however, add our sincerely meant (but rather traditional) hope that it will prove interesting to graduate students, to colleagues and to anyone else, who will bother to read it.

The book was written as a joint effort by a theoretically inclined population geneticist and an experimental ecologist who share opinions on what is interesting in the field of theoretical ecology. While we believe that qualified natural history is of indisputable intrinsic value, we think that ecology is a natural science which should have a theoretical framework. On the other hand, theoretical ecology must draw its inspiration from nature and yield results which give insight into the findings of the naturalist and inspire him to make new observations and experiments. Without this relationship between field biology and theory, mathematical ecology may become a discipline totally divorced from biology and solve—albeit interesting—mathematical problems without significance for ecology. Therefore, in addition to theoretical population biology (including some original models) the book also discusses observational data from nature to show how the theoretical models give new insight and how observations give rise to new theoretical thought. While no book on ecology could do without the mention of the hare-lynx example (and ours is, therefore, no exception) we have tried to bring new examples mainly derived from one of the authors' field of experience: microbial ecology and marine biology.

We are aware that the book does not cover all important or even interesting aspects of theoretical ecology. In addition to this lack of completeness we might add two other aspects not covered. We do not treat realistic population models of particular systems. While such models are important, not least from a practical point of view (e.g. pest control, fishery science and forestry) we do not feel that models with little generality belong within the framework of this book. The book does not treat what is usually referred to as "systems ecology", a subject which more often than not has provided excuses for collecting large numbers of trivial data and for substituting insight with computer outputs which resemble the above-mentioned data.

Ecology is a relatively new science and it deals with very complex systems. It has, therefore, sometimes suffered from lack of stringency on the part of its students. This situation has not been improved by the recent tendency of ecology to become a kind of public ideology. The question of stability and diversity, thoroughly discussed and analyzed by May (1973a), examplifies this. What May calls the "conventional > complexity begets stability < wisdom" has in many minds ideological undertones implying a special kind of perfection of nature and how man spoils it all. The word "strategy" is usually a harmless synonym for adaptation when used in connection with single species; the term "ecosystem strategy" is a meaningless phrase explaining nothing (as if the members of an ecosystem by mutual agreement decided to ration phosphorus or to feed the pill to "$K$-strategists" in order to retain community stability). We have throughout the book, to the best of our ability, tried to be rigorous in our choice of terms. We believe that properties of biological communities should be explained as the result of the properties of single populations shaped by natural selection and that selection acts through individual fitness.

One of our original objectives was, if not to integrate, then to demonstrate the connection between population genetics and population ecology. This idea is not new, and as with previous attempts we have not been fully successful. Most elementary textbooks which attempt to integrate the two subjects have in fact resulted in collecting separate chapters on the two fields within one book. With respect to experimental and field studies of populations, there are unfortunately not many examples to draw on yet. The school of "ecological genetics" is mostly genetics and the ecology is not very sophisticated. Most contributions to the unification of genetics and ecology by other workers consist mainly of studies on isozyme patterns in relation to environmental gradients; the genetical substance of these studies is rarely significant. As an example of a recent successful attempt to study ecological genetics in the field, the reader is referred to Volume 6 in this series (Stern and Roche, 1974).

On the theoretical side many ecologists have recognized the evolutionary significance of ecology and thus the connection to genetics. However, their models are usually devoid of sex, that is their alleles or genotypes behave like distinct species in models of competition. So Mendel and Hardy-Weinberg are excluded and with them the whole basis of population genetics (and the realism of the models as well when treating outbreeding species). Population geneticists, on the other hand, have largely ignored absolute population sizes in their models. Since ecologists devote most of their lives to studying population sizes, the starting point has not been too hopeful. Recently, however, several workers have contributed to the unification of population genetics and ecology with respect to the theory of density dependent selection and the evolution of the niche; this work is discussed together with a few original contributions in Chapters 1 and 3.

The structure of the book is as follows. In Chapter 1 we discuss aspects of population growth and fitness and we explore different models of density-dependent growth and density-dependent selection. Finally we give a critical discussion on how the concept of "$r$- and $K$-selection" has been used for describing the properties of real populations.

In Chapter 2 we explore different models of interspecific interactions and discuss the stability of multispecies systems with competitive, mutualistic and prey-predator interactions.

In Chapter 3 we discuss the concept of the niche and limiting similarity. This chapter is written with the conviction that the theory of niche packing and limiting similarity is one of the most important achievements of theoretical ecology and we discuss evidence from nature in some detail.

In Chapter 4 we discuss models of geographically structured populations and the significance of migration-extinction equilibria. The last section of the chapter is devoted to the theory of island biogeography which, together with the theory of limiting similarity, constitutes the most important component of a unifying theory of ecology so far achieved.

Finally, in Chapter 5, we try to bring the theories discussed in this book in context with a spectrum of other ideas and theories pertaining to the community concept.

Within some of the chapters a few sections only describe the mathematical derivation of some expressions. Also in some places mathematical techniques usually not mastered by biologists are applied. Such sections are printed with small type; they are not necessary for a general understanding of the book or the models they describe.

*Acknowledgements.* The idea to write this book originated after a series of seminars on theoretical population biology held at the Department of Ecology and Genetics, University of Aarhus in 1974–75. These seminars were attended not only by ecologists and geneticists but also by mathematicians and statisticians. Our gratitude is due to all the participants for inspiring discussions and for single contributions on various topics. We want in particular to express our gratitude to the following who contributed with ideas, references and discussions on particular topics treated in this book: Paul Harrison (the history of the concept of the niche), Jørgen Hylleberg (marine bottom communities), Niels Keiding (stochastic extinction of populations), Hans H. Lassen (the theory of island biogeography), and Ebbe Thue Poulsen (the model on density dependent growth described in Chapter 1; this model will also be published separately). Timothy Prout contributed specifically in clarifying our ideas on density dependent selection; his many inspiring ideas and comments, however, have influenced our thinking on nearly all the topics treated in this book.

One of us (F.B.C.) originally took interest in theoretical ecology by attending a series of seminars held by Jonathan Roughgarden at Stanford; special gratitude for this inspiration is expressed here.

Finally we want to acknowledge the careful reviews of this book by Werner Fenchel and by Niels Keiding who helped to clarify our minds on many of the subjects covered. A reading of the final draft by J. Stuart F. Barker improved the English considerably.

Aarhus, January 1977

FREDDY B. CHRISTIANSEN
TOM M. FENCHEL

# Contents

# 1. Population Growth and Fitness

The principal variable in any population ecological model is the population size(s) of the organism(s) under investigation within the habitat or geographical area of interest. The size of a specific population may be measured in numerous units which are more or less equivalent. An obvious unit to consider is the individual organism, and the population size is then expressed as the number of individuals in the population or as the density of individuals within the area of investigation. This approach is well suited for the description of populations of most animals, whereas individual counts may cause problems in some species of plants. Alternatively, we may measure the population size in units of biomass, i.e., the weight or energy content.

These two ways of keeping account of the population will in many organisms be equivalent: given the size distribution of the individuals in the population, transformations from number of individuals to biomass and back are possible. In the following, we will use the general term size, and by that refer to either of the two basic measurements or to transformations thereof.

In this chapter we will present a discussion of the simplest models which describe the growth and regulation of single species populations without reference to the species possibly interacting with other species in the habitat. We will also discuss some simple models allowing the basic parameters for the growth and regulation of the population to be shaped by natural selection.

## 1.1 Density-Regulated Population Growth

A theory for population growth and regulation was essentially initiated by the observation by Malthus (1798) that if one considers population growth as a result of the reproduction characteristics of individuals, then populations either grow indefinitely with a constant doubling time or they vanish. According to Malthus the only possibility for a constant or even for a limited, nonzero population size is that the initial population size is conserved through time. The rationale behind Malthus' model is that the individuals of the population are considered to be identical and independent and they are considered to die off at a constant rate $d$ and to reproduce at a constant rate $b$. If the population is assumed to be so large that the population size may be measured by a continuous variable $x(t)$ through time, then the model may be stated.

$$dx/dt = (b-d)x \tag{1.1}$$

which integrates to

$$x(t)=x_0 \exp[(b-d)t] \tag{1.2}$$

where $x_0$ is the initial population size at time zero. The birth and death rates are connected in the model such that the population behavior is completely described by the parameter $r=b-d$; $r$ is referred to as the Malthusian parameter, the intrinsic rate of increase, the *reproduction potential* or the unlimited growth rate. From Eq. (1.2) it appears that the population grows infinitely if $r>0$, decreases to a vanishingly small size if $r<0$, and conserves the size $x_0$ for $r=0$. Neither of these results are general characteristics of natural populations, and at least the unlimited growth is obviously impossible for any extended period of time. The growth of natural populations is thus inadequately described by the model.

Malthus' model of population growth has been the reference model of demography; extended analysis has shown that the prediction of the model rests on the assumption of independent life and reproduction of the individuals. Thus, population limitation and stability must be the result of inhibitory interactions between the individuals in the population at high densities.

The simplest possible model of density regulation is the logistic model for population growth (Verhulst, 1838; Pearl and Reed, 1920). The model assumes that the reproductive rate per individual decreases linearly with population size, i.e.,

$$dx/dt=rx(K-x)/K\,. \tag{1.3}$$

The two constants may be interpreted as (1) the reproduction potential, $r$ (assumed positive), of an individual in the absence of inhibition from other individuals, i.e., the population growth rate at low densities, and (2) the *carrying capacity*, $K$, of the environment, i.e., the upper limit of the population size that allows growth.

Equation (1.3) integrates by observing that

$$d(1/x-1/K)/dt(=-x^{-2}dx/dt)=-r(1/x-1/K)$$

and, therefore,

$$1/x-1/K=(1/x_0-1/K)\exp(-rt)$$

where $x_0$ is the population size at time zero. This rearranges to

$$x(t)=[Kx_0 \exp(rt)]/[(K-x_0)+x_0 \exp(rt)]\,. \tag{1.4}$$

For relatively small initial population sizes (i.e., small $x_0/K$) we get approximately exponential growth, i.e.,

$$x \approx x_0 \exp(rt)$$

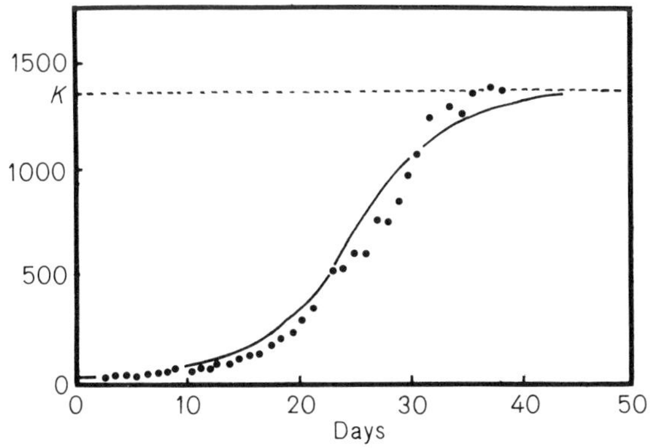


Fig. 1. The growth of a population of *Hydra* in a mixed culture of freshwater organisms and a logistic curve fitted to the data. (Redrawn from Bick, 1964)

and after some time we get convergence of $x$ to $K$ at an exponential rate, i.e.,

$$|K-x| \approx (\text{const.}) \exp(-rt)$$

(see Fig. 1). The population will thus stabilize at the size $K$.

The model as a description of a natural population, rests on a range of assumptions about the life and death of the individuals and about the interactions between the individuals. Among these assumptions may be mentioned the following:

1. The response in the multiplication rate as function of population size is instantaneous, thus no time-lags are allowed (e.g. as the time from birth to reproductive age, or other effects of age structure in the population).

2. The carrying capacity is independent of the population size in the past; a resource cannot be eaten up.

3. The interactions between the individuals are described solely by an effect of filling up limited space; no effect like increased reproductive success of sexual organisms with increasing population size (i.e., an Allee effect), or alternatively antagonistic interactions which are proportional to random encounters of individuals are considered in the model.

This last assumption of the model [Eq.(1.3)] may have only restricted implications with respect to its qualitative predictions. Assuming a real population for which $dx/dt \geqq 0$ for $x$ less than a value $K>0$ and negative elsewhere, the logistic model will still give a pretty good description of the population growth over a limited range of population sizes. The model may even be used in connection with an Allee effect, i.e., that the population cannot increase at low densities, if we restrict our attention to population sizes above a certain level.

A correlation between resources and population sizes in the past [i.e., 2. discussed above] may be built into the model, for instance

$$dx/dt = rx(t)[K - x(t-\tau)]/K \tag{1.5}$$

as discussed by Hutchinson (1948). Such models may be used to describe Nicholson's (1954) population cage experiments with blow flies (see Fig. 3). In one of his experiments Nicholson supplied a constant amount of food per day to the adult flies and unlimited amounts of food to the larvae. The egglaying of the flies is limited by the food availability, such that a high adult population density results in a low production of eggs and a low density of adults results in a high egg production. In the model [Eq. (1.5)] the population may produce considerable overshoots relative to the carrying capacity resulting in regular oscillations in population size through time. Interpretations of Nicholson's (1954) data have been thoroughly discussed by May (1973a), Maynard Smith (1974), and Oster (pers. comm.).

An alternative, and probably more realistic formulation of the effect of a time-lag is

$$dx/dt = rx(t)\left[K - \int_0^\infty f(\tau)x(t-\tau)d\tau\right]\Big/ K \tag{1.6}$$

where $f$ is some weighting distribution. The analysis of this kind of model has not proceeded very far, but presumably Eqs. (1.3) and (1.5) may be considered to delimit the possibilities. These time-lag models of resource-population interactions share many properties with the time-lag models constructed in order to describe deviations from assumption 1. of the logistic model. An important class of such models are the discrete generation models which will be discussed later in this section.

The parameters of the logistic model [Eq. (1.3)], i.e., the reproduction potential and the carrying capacity, may be deduced from suitable assumptions on the density dependence of the birth and death rates in a way parallel to the construction of the Malthusian model. Suppose that the birthrate of the population is $b$ at all densities and that the death rate increases linearly with population size, i.e., equals $d + \iota x$, where $d$ is the basic death rate of a small population where the individuals do not interact and $\iota$ is a measure of the interaction between individuals. According to this model, the growth of the population is described by

$$dx/dt = x[b - (d + \iota x)]\,. \tag{1.7}$$

By rearranging the terms in Eq. (1.7) we see that it corresponds to the logistic model [Eq. (1.3)] with $r = b - d$ (as in the Malthusian model) and $K = (b - d)/\iota$. This formulation provides a more plastic definition of the carrying capacity $K$ and it might be used as a tool for predicting the change in the equilibrium population size following changes in the environment. A deterioration of the environment of the organism could be modeled as a relative increase in the basic death rate: $d' = d(1 + \varepsilon)$. The effects of this change are to decrease the reproduction potential $r$ and the carrying capacity $K$, which per definition is the equilibrium population size. This decrease in $K$ is a reflection of a new balance between birthrate, death rate, and the intensity of the interaction. However, the change is in this model independent of the birthrate, viz.,

$$K' - K = -d\varepsilon/\iota\,. \tag{1.8}$$

Many organisms are characterized by regular breeding cycles which correspond to different seasons of the year. The growth of such populations is often well described by a continuous approximation like Eq. (1.3) if considered over a long period of time. This kind of approximation is parallel to the description of the population size as a continuous variable. However, a closer description of the growth and regulation of populations of such species is of course expected through the application of similar simple discrete generation models. For simplicity, let us confine our attention to the simplest situation of discrete generation models where the interval between breedings corresponds to the length of life of the individuals. These models may apply to the description of populations of, e.g., annual plants or insects. These simple models are commonly referred to as *discrete nonoverlapping generation models*.

Another reason for considering such models is that they are more useful for studying single components of the life cycle of organisms. From an evolutionary point of view, breeding is an important event in the life of an individual which involves mating and segregation. These processes simplify considerably in models with discrete nonoverlapping generations; we will return to this aspect of the models later.

We will here consider two discrete generation models which are obtained by arguments parallel to those leading to the two different parameterizations of the logistic model as represented by Eqs. (1.3) and (1.7) respectively. Anderson (1971) and Roughgarden (1971) considered the following model,

$$\Delta x = Rx(K-x)/K\,, \tag{1.9}$$

where $\Delta x$ is the change in population size between two generations, $R>0$ is the growth rate per individual per generation of the population during one generation for a small population with no interaction between the individuals, and $K$ is the carrying capacity. An alternative and often convenient formulation of Eq. (1.9) is in terms of the recurrence equation

$$x' = x[(1+R)-(R/K)x] \tag{1.10}$$

where $x'$ is the population size of a generation following a generation with the population size $x$.

From Eq. (1.9) it is inferred that the population size is increased in successive generations if $x<K$, decreased if $x>K$, and unchanged if $x=K$. In order to study the behavior of the population around the equilibrium $\hat{x}=K$, consider Eq. (1.10) when the population size deviates a little from the equilibrium, i.e., $x=K+e$. With this notation it follows that

$$e' = e(1-R)-(R/K)e^2\,. \tag{1.11}$$

If the population is close to equilibrium, i.e., $e$ is close to zero, then the first term of the recurrence Eq. (1.11) will dominate. Therefore, if $0<R<1$, then $e'$ will have the same sign as $e$ and $|e'|<|e|$ such that the population will approach the equilibrium without passing it. On the other hand, if $R>1$, then $e'$ and $e$ will

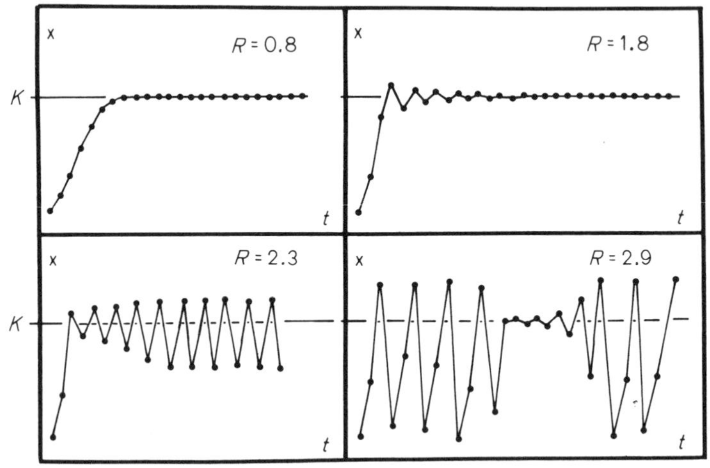


Fig. 2. The behavior of populations growing according to a discrete generation model [Eq. (1.9)] for different values of $R$

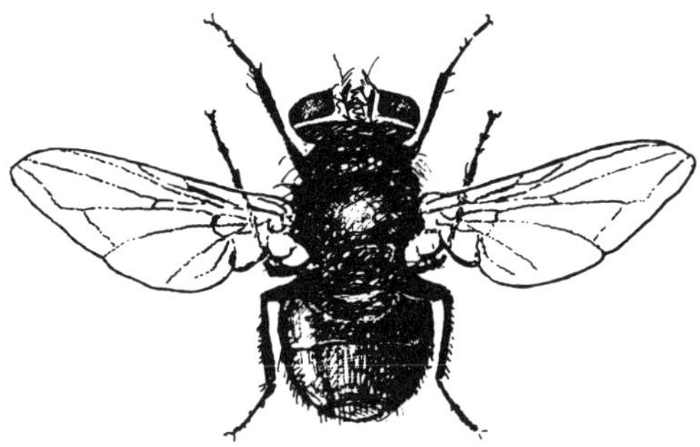

Fig. 3. A blowfly

have opposite signs and a population with a size above but close to $K$ will in the next generation have a population size smaller than $K$. Furthermore, when $R>2$, then $|e'|>|e|$ for all $e$ sufficiently small such that the deviation from the equilibrium will increase in size and the population will oscillate away from the equilibrium. For $1<R<2$, on the other hand, the population will show damped oscillations around the equilibrium, approaching it closer for each generation (Fig. 2).

From Eq. (1.10) it is seen that the model is only a reasonable description as long as the initial population size $x$ is restricted to the interval

$$0<x<(1+R)/(R/K) \tag{1.12}$$

because otherwise we would get $x' \leqq 0$, i.e., the population would collapse immediately. However, the further requirement that $x'$ should belong to this interval when $x$ does is equivalent to $R<3$. Thus when $2<R<3$, a population started in the interval [Eq.(1.12)] will stay there and perform oscillations between high and low population sizes, either as a stable limit cycle or as complete chaos (Fig. 2).

The possibility for oscillations in the model [Eq.(1.10)] is due to its inherent time-lag (see May, 1975). The multiplication in population numbers during reproduction at the end of a generation period depends only on the size of the population as newborns, a situation which resembles that of Nicholson's (1954) experiments as discussed in connection with the model described by Eq.(1.5). However, as long as we restrict the use of the model to situations where $R$ is small, then its behavior is rather close to that of the continuous time logistic model and we may take advantage of the simple mathematical formulation of Eq.(1.10).

The second discrete generation model, formulated by the same principle as the model [Eq.(1.7)], is due to Poulsen (1975). The key observation is that discrete generations mean discrete breeding and not necessarily a discrete death event. The change in population size, $u$ say, during a generation between two successive periods of reproduction is described by the equation

$$du/dt = -u(d + \iota u) \tag{1.13}$$

from $t=0$ to the generation time $T$, such that $u(0)=Bx$, where $B$ is the number of offspring produced by each adult individual. The recurrence equation between two generations is then

$$x' = u(T)\,. \tag{1.14}$$

By observing that Eq.(1.13) is just another version of Eq.(1.3) we get from Eq.(1.4) that

$$u(t) = (Bx\exp(-td))/[1 + \delta xB(1 - \exp(-td))]\,, \tag{1.15}$$

where $\delta = \iota/d$ is the interaction constant relative to the basic death rate. From this and Eq.(1.14) we get the recurrence equation for the model as

$$x' = (xBD)/[1 + \delta xB(1 - D)] \tag{1.16}$$

where we have put $D = \exp(-Td)$ as a basic, discrete death rate parameter.

It can be seen from Eq. (1.16) that when $BD < 1$, the population will decrease through time and eventually vanish. When $BD > 1$, then the population size will increase when low but decrease when sufficiently high. A stationary point is found as the solution to the equation $x' = x = \hat{x}$ which from Eq.(1.16) gives

$$\hat{x} = (BD - 1)/[\delta B(1 - D)]\,: \tag{1.17}$$

below this point the population size will always increase and above it it will always decrease. This equilibrium point corresponds to the carrying capacity, $K$, of the environment. In order to study the behavior around the equilibrium point

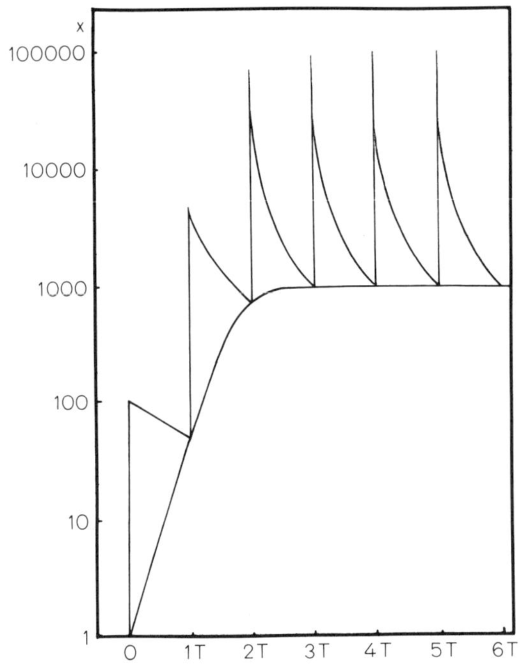


Fig. 4. A population growing according to model [Eqs. (1.13) and (1.14)]. The smooth curve connects the sizes of the adult population at the end of each generation. (Redrawn from Poulsen, 1975)

write recurrence Eq. (1.16) in terms of the population size with a small deviation from equilibrium, i.e., $x = \hat{x} + e$;

$$e' = e/[BD + e\delta B(1-D)] . \tag{1.18}$$

When close to the equilibrium ($e$ small) the first term in the denominator dominates so that the multiplication factor of $e$ between generations is approximately $1/BD$ which is less than unity whenever the equilibrium [Eq. (1.17)] exists. Therefore, the deviation will decrease for each generation and the signs of $e$ and $e'$ are the same so that the equilibrium will be approached without passing it.

The properties of this model are thus very parallel to what was found for the continuous time logistic model. This close relation becomes even more apparent if in Eq. (1.16) we substitute $K = \hat{x}$ as given by Eq. (1.17). This yields

$$x' = x(BD)K/[(K-x) + xBD] \tag{1.19}$$

which is of the form of Eq. (1.4) with $BD = \exp(rT)$ where $T$ is the generation length. Actually the time dependent behavior is

$$x^{(n)} = x(BD)^n K/[(K-x) + x(BD)^n] \tag{1.20}$$

where $x^{(n)}$ is the population size after $n$ generations. Thus the model [Eq.(1.16)] is really the discrete time analog of the logistic model. The behavior of a population growing according to the model is shown on Figure 4.

This model provides an even more plastic definition of the carrying capacity $K$ than does the parameterization [Eq.(1.7)] of the logistic model. The density limitation of the population numbers does not occur at population sizes around the carrying capacity but rather in populations of young individuals at very high densities (Poulsen, 1975). This remark pertains of course only to the adult population; if we choose to count the population of newborn individuals, then their carrying capacity is $BK$; at equilibrium they "hit the ceiling" in each generation and the population limitation occurs at population densities around the carrying capacity.

## 1.2 Genetical Models of Population Growth

When simple models of population growth are considered it is assumed that all the individuals in a population are identical and unchangeable. However, natural populations are characterized by a high degree of genetic variation (see, e.g. Dobzhansky, 1970; Lewontin, 1974) which makes the individuals different and changeable. From the ecologist's point of view the primary interest is the states and changes in the variation that directly influence the relation between the population and its environment. The change in the genetic composition of the population through time is explored in the science of population genetics; in the following we will introduce some concepts from this field and discuss some problems on the borderline between genetics and ecology.

A fundamental consideration in population genetics is the situation of two alternative alleles at a locus in a diploid and outbreeding organism. Denote the two alleles $A$ and $a$ and suppose that the three genotypes $AA$, $Aa$ and $aa$ occur in the numbers $x_1, x_2$, and $x_3$ respectively in the population so that the total population size is $x = x_1 + x_2 + x_3$. In order to facilitate the formulations we introduce two additional sets of parameters, the genotypic frequencies $f_i = x_i/x$ $(f_1 + f_2 + f_3 = 1)$ and the frequencies of $A$:

$$p = (2x_1 + x_2)/(2x) = f_1 + (1/2) f_2$$

and of $a$: (1.21)

$$q = (x_2 + 2x_3)/(2x) = (1/2) f_2 + f_3 = 1 - p\,.$$

The gene frequencies are simply the frequencies of the alleles in a pool of gametes if all individuals contribute the same number of gametes to this pool. We will now and in the following assume that populations reproduce by random union of gametes. This may occur by random pair formation (random mating) and equal fertility of all pair formations. In this situation a new zygote is formed by the union of two gametes, one from each parent, and when the parents are chosen at random, then we might as well choose the two gametes at random from all gametes in the population. A probably more obvious example of random union of gametes is found in marine organisms which show synchronized spawning. If we denote the gene frequencies in the male gametes $p_{♂}$ and $q_{♂}$ and in the female gametes $p_{♀}$ and $q_{♀}$, then by random union of the gametes the frequency of the genotype $AA$ among newly formed zygotes becomes $p_{♀}p_{♂}$. Similarly the frequency of the $Aa$ genotype becomes $p_{♀}q_{♂}+p_{♂}q_{♀}$ and that of the $aa$ genotypes becomes $q_{♀}q_{♂}$. If we in addition to random union of gametes also assume that the gene frequencies in males and females are equal, i.e., $p_{♀}=p_{♂}=p$ and $q_{♀}=q_{♂}=q$, then the proportions of the genotypes in the newly formed zygotes become $p^2$, $2pq$, and $q^2$ for $AA$, $Aa$ and $aa$ respectively. These proportions are the Hardy-Weinberg proportions. Note that the zygote formation by random union of gametes does not alter the gene frequency, since $p^2+1/2(2pq)=p(p+q)=p$.

From these results we get the internal structure of the zygote population; however, we will have to determine its contribution to the population, from our ecological model. In discrete population models, the zygote population initiates a new generation and all we have to determine is the size of the zygote population $u(0)$ which then determines the exact composition of the population as

$$u_1(0)=p^2u(0),\ u_2(0)=2pq(0),\ u_3(0)=q^2u(0), \tag{1.22}$$

In the discrete time logistic model, the Poulsen model (given by Eq. (1.13) and Eq. (1.14) the change in the population size and composition is now described by the equation,

$$du_i/dt=-u_i(d+\iota u)\,. \tag{1.23}$$

With this model and with no difference between the genotypes the genotypic proportions do not change during the generation and we can thus expect the genotypes to occur in Hardy-Weinberg proportions throughout the generation.

If we consider a continuously breeding population (such as the logistic model modeled by Eq. (1.7)) the rate of production of zygotes is $bx$. The zygotes are in Hardy-Weinberg proportions such that the population will evolve according to

$$\begin{aligned} dx_1/dt&=p^2xb-x_1(d+\iota x)\,,\\ dx_2/dt&=2pqxb-x_2(d+\iota x)\,,\\ dx_3/dt&=q^2xb-x_3(d+\iota x)\,, \end{aligned} \tag{1.24}$$

if there are no differences between the genotypes in relation to growth and reproduction. The total population grows according to Eq. (1.7) which is obtained by summing Eq. (1.24). The gene frequency does not change in this model ($dp/dt=0$) and if the population is in Hardy-Weinberg proportions (e.g., $x_1=p^2x$) then it will remain so ever after ($df_i/dt=0$). Furthermore, if the population deviates from these proportions initially, then they will converge towards them.

This can be shown by considering the commonly used measure of deviations from Hardy-Weinberg proportions

$$F=(4f_1f_3-f_2^2)/(4pq)\,. \tag{1.25}$$

which is such that $f_1=p^2+pqF$, $f_2=2pq(1-F)$, and $f_3=q^2+pqF$. Differentiating Eq. (1.25) yields (after some calculations) the equation

$$dF/dt=-bF\,,$$

which integrates to

$$F(t)=F(0)\exp(-bt)\,. \tag{1.26}$$

This means that deviations from the Hardy-Weinberg proportions decrease at an exponential rate which is the birthrate. Thus, the genotype frequencies converge to the Hardy-Weinberg proportions considerably faster than the rate $(b-d)$ at which the population converge to its equilibrium size. We will, therefore, expect that a population at equilibrium size will be in Hardy-Weinberg proportions at a locus where the genotypes do not differ with respect to the population growth parameters.

These laws of the Hardy-Weinberg proportions constitute the basic laws of population genetics but in order to make the theory interesting to ecology we must suppose that the genotypes are different with respect to the population parameters. In the model [Eq. (1.24)] even simple differences between the genotypes yield a mathematically complicated model. This is in part due to the fact that the Hardy-Weinberg law is not absolute but the convergence to it is gradual as shown by Eq. (1.26). In discrete generation models, however, each generation is initiated by a population structure as shown in Eq. (1.22); we have a simple recurrent structure in the model since the three dimensional variable $(u_1, u_2, u_3)$ is determined by the two dimensional variable $(p, u)$.

First let us explore the Poulsen model given by Eq. (1.13) and Eq. (1.14). The first observation to make is that Eq. (1.23) describing the time-dependent behavior of the population between two breeding periods has not been integrated in the situation where $d$ and $\iota$ depend on the genotype. Thus, we will initially restrict our attention to models where the differences between the genotypes show either before or after the time of the density-dependent death, i.e., models which are basically of the same structure as the commonly used population genetics models. A simple *density-independent* selection model was considered by Poulsen (1975). In this model the zygotes of the different genotypes have different probabilities of initiating their development (e.g., the probability that eggs will hatch). With this model we get instead of Eq. (1.22) as the initial population composition the following:

$$\begin{aligned} u_1(0)&=P_1p^2(Bx)\,,\\ u_2(0)&=P_2 2pq(Bx)\,,\\ u_3(0)&=P_3q^2(Bx)\,, \end{aligned} \tag{1.27}$$

where $P_i$ is the survival probability of the zygotes. From Eq. (1.27) we get by summation

$$u(0)=\bar{P}(Bx)$$

where

$$\bar{P}=P_1p^2+P_2 2pq+P_3q^2 .$$

The growth of the population Eq. (1.27) is governed by Eq. (1.23). Thus from Eq. (1.14) and Eq. (1.16) we get the composition of the adult population

$$\begin{aligned} x_1' &= P_1p^2xBD/[1+\delta\bar{P}xB(1-D)] ,\\ x_2' &= P_2 2pqxBD/[1+\delta\bar{P}xB(1-D)] ,\\ x_3' &= P_3q^2xBD/[1+\delta\bar{P}xB(1-D)] , \end{aligned} \tag{1.28}$$

with the total population size

$$x'=\bar{P}xBD/[1+\delta\bar{P}xB(1-D)] , \tag{1.29}$$

and where $\delta=\iota/d$.

Comparing Eq. (1.28) and Eq. (1.29) reveals that the gene frequency has changed independently of the initial population size to

$$p'=p(P_1p+P_2q)/\bar{P} , \tag{1.30}$$

with the change in gene frequency

$$p'-p=pq[p(P_1-P_2)+q(P_2-P_3)]/\bar{P} . \tag{1.31}$$

The change in gene frequency is zero if either $p=0$ or $q=0$ or if

$$p=\hat{p}=(P_2-P_3)/(2P_2-P_1-P_3) , \tag{1.32}$$

which are the three potential equilibrium points. We only have to consider $\hat{p}$ as an equilibrium point if it is positive and less than unity. So Eq. (1.32) is only a valid equilibrium point if either $P_2>P_1$ and $P_3$ or if $P_2<P_1$ and $P_3$. The first possibility where the survival probability of the heterozygote is larger than that of both homozygotes is called *overdominance* (or heterosis) and the second possibility where the survival probability of the heterozygote is less than that of both homozygotes is called *underdominance.* The additional cases will be called *directional selection*; when $P_1 \geqq P_2 \geqq P_3$ then the square bracket in Eq. (1.31) is always positive, i.e., $p'>p$ and the gene frequency of $A$ will always increase until it is fixed in the population and $a$ is absent. Similarly, when $P_1 \leqq P_2 \leqq P_3$ then we have directional selection in favor of $a$ which will eventually become fixed in the population.

When we have over- or underdominance, then the square bracket in Eq. (1.31) has different sign according to whether $p<\hat{p}$ or $p>\hat{p}$. In the case of overdominance the sign is positive when $p$ is below the equilibrium point and negative when $p$ is above the equilibrium point; thus in this case the gene frequency will increase below and decrease above the equilibrium. If we further observe that

$$p'-\hat{p}=(p'-p)+(p-\hat{p})=[1-pq(2P_2-P_1-P_3)/\bar{P}]\,(p-\hat{p})<(p-\hat{p})\,,$$

we see that the gene frequency can never pass the equilibrium. Therefore, in the case of overdominance the gene frequency will converge towards the equilibrium gene frequency $\hat{p}$ given by Eq. (1.32).

For underdominance the situation is reversed; the gene frequency will increase above the equilibrium point and decrease below it. Depending on whether the population starts above or below the equilibrium one of the genes will become fixed and the equilibrium point $\hat{p}$ is therefore unstable.

In any of the selection cases the change in gene frequency is always so as to increase the mean survival probability $\bar{P}$. This may be argued intuitively by differentiating $\bar{P}$ with respect to $p$ and remembering that $q=1-p$:

$$\partial\bar{P}/\partial p=2P_1p+2P_2(q-p)-2P_3q\,, \tag{1.33}$$

which inserted into Eq. (1.31) reveals that

$$p'-p=[pq/(2\bar{P})]\,\partial\bar{P}/\partial p \tag{1.34}$$

Thus, from points where $\bar{P}$ is an increasing function of $p$, the gene frequency will be increasing, and from points where $\bar{P}$ is a decreasing function of $p$ the gene frequency will be decreasing. The equilibrium point $\hat{p}$ in Eq. (1.32) will in the case of overdominance be the maximum of $\bar{P}$ and in the case of underdominance it will be the minimum.

After this digression into population genetics we can return to the ecological problem of the influence of selection on population growth and regulation. The differential survival of the early zygotes modifies the number of offspring per parent, $B$, as seen in Eq. (1.29). Therefore, the condition for increase in the population size now becomes $\bar{P}BD>1$. Under this condition, as seen from Eq. (1.29), we see that the population size changes towards

$$K(p)=(\bar{P}BD-1)/[\delta\,\bar{P}B(1-D)]\,. \tag{1.35}$$

We may even consider this expression as a measure of the population size through the selection process if the change in gene frequency is very slow compared to the rate of convergence of the population size to equilibrium. The dependence of $K(p)$ on the gene frequency is illustrated by the relation

$$[K(p_2)-K(p_1)]/K(p_1)=\{[\bar{P}(p_2)-\bar{P}(p_1)]/\bar{P}(p_2)\}/[\bar{P}(p_1)\,BD-1]\,. \tag{1.36}$$

From this it is seen that the effect of the gene frequency $p$ on $K(p)$ may be strong if $\bar{P}BD$ is close to unity, whereas if $B$ is very large, then Eq. (1.36) may be as close to zero as we wish. Thus, dependent on whether $B$ is small or large, the adult population size will show dependence or virtual independence of the gene frequency.

Another kind of density-independent selection model that fits into the framework of the discrete generation logistic model is one in which differences in the fertility of the different genotypes are allowed (Prout, pers. comm.). Assume that the proportions $F_1$, $F_2$, and $F_3$ of the adult genotypes are sterile, or for some other reason do not participate in breeding. Then the effective adult population is $x_i^e = F_i x_i$, such that

$$x^e = \bar{F}x, \quad \text{with} \quad \bar{F} = F_1 f_1 + F_2 f_2 + F_3 f_3 \tag{1.37}$$

where $x^e$ is the number of breeding adults and $f_i$ is the frequency of genotype $i$ among all adults.

If we assume that the adult population is in Hardy-Weinberg proportions with gene frequency $p$, then from Eq. (1.21) the gene frequency of the gametes is

$$p' = p(F_1 p + F_2 q)/\bar{F} \tag{1.38}$$

and the composition of the zygote population is now given by

$$\begin{aligned} u_1(0) &= (p')^2 B\bar{F}x\,, \\ u_2(0) &= 2(p'q')B\bar{F}x\,, \\ u_3(0) &= (q')^2 B\bar{F}x\,. \end{aligned} \tag{1.39}$$

The population will thus be in the Hardy-Weinberg proportions throughout the generation, thus justifying the previous assumption. Comparing this with Eq. (1.27) we see that apart from the genotypic proportions, the two models are equivalent ecologically speaking. In both instances, the net effect of genetic variation is to change the net birth factor $B$. From a genetical point of view, however, the difference is large since in the model [Eq. (1.39)] the population is in Hardy-Weinberg proportions whenever observation is possible whereas Eq. (1.27) does not fit these proportions [it may if the survival probabilities have a special structure, so existence of the Hardy-Weinberg proportions is not a certain criterion for the absence of selection (Prout, 1965)]. The fecundity model considered here has an interesting counterpart in sexual organisms which mate. If the fecundity of a pair is determined by the female only and if all females are mated, then even strong selection for mating selectivity in males will have no effect on the population size of the zygotes. This kind of selection which does not influence population size has been termed soft selection by Wallace (1968). However, if the birth factor $B$ is very large, then the zygotic selection model [Eq. (1.27)] is equivalent to a soft selection model, and so is the model [Eq. (1.39)].

The models considered so far are the commonly used models of population genetics, that is models of density independent selection. More elaborate discussions of such models may be found in population genetics textbooks (e.g. Ewens, 1969; Crow and Kimura, 1970). The additional feature of the models which may be explored in an ecological context is the influence of density-independent selection on population size in a density-regulated population (Poulsen, 1975; Prout, in preparation).

We will now turn to the problem of *density-dependent* selection. We have already mentioned that when the three genotypes have different death parameters ($d$ or $\iota$) we cannot integrate Eq. (1.23) to get the recurrence equations for gene frequencies and population sizes. However, we can apply a technique which is commonly used in genetics; viz. to find the conditions for the initial increase of rare genes. Suppose that gene $A$ is very rare, then the homozygote $AA$ will be extremely rare and in the population the gene $A$ will predominantly be found in heterozygous individuals, $Aa$. Thus, in this state, the population may be considered to consist of only two types of individuals, viz. $Aa$ and $aa$ which occur in the numbers $2pBx$ and $(1-2p)Bx$ respectively, if we neglect quantities of the order $p^2$, where $p$ as before is the gene frequency of $A$. Also assuming as above that the zygotes have different probabilities of initiating their development, then we get

$$u_2(0)=2pP_2Bx, \quad u_3(0)=(1-2p)P_3Bx\,. \tag{1.40}$$

From this stage the decrease of the zygote population from the initial composition and size [Eq. (1.40)] to the adult stage is determined by Eq. (1.23), which with the assumption of genotype-dependent death parameters, reads

$$\begin{aligned} du_2/dt &= -u_2(d_2+\iota_2 u)\,, \\ du_3/dt &= -u_3(d_3+\iota_3 u)\,. \end{aligned} \tag{1.41}$$

This system cannot be solved explicitely, but the solution $[u_2(t), u_3(t)]$ is continuously differentiable in the initial values $[u_2(0), u_3(0)]$, and the initial values are differentiable functions of $p$, such that the solution to Eq. (1.41) is a differentiable function of $p$. To emphasize this we will for the present calculations write the solution of Eq. (1.41) as $[u_2(t,p), u_3(t,p)]$. Our interest is to express the gene frequency after one generation, $p'$, as a function of the initial gene frequency $p$ under the assumption that $p$ is small. By Taylor expansion of $p'$ we get using $u_2(t,0)=0$:

$$\begin{aligned} p' &= u_2(T,p)/\{2[u_2(T,p)+u_3(T,p)]\} \\ &= \{\partial u_2/\partial p(T,0)/[2u_3(T,0)]\}\; p + \text{terms of order } p^2\,. \end{aligned} \tag{1.42}$$

Thus, the condition for increase of the frequency of $A$ when rare is that the factor of $p$ in Eq. (1.42) should be larger than unity. In the following this condition is calculated, with the result shown as Eq. (1.47).

From Eq. (1.41) we have, by again using $\delta_i = \iota_i/d_i$,

$$\frac{d}{dt}\left(\frac{\partial u_2(t,p)}{\partial p}\right) = \frac{\partial}{\partial p}\left(\frac{du_2}{dt}\right) = -d_2(1+\delta_2 u)\frac{\partial u_2}{\partial p} - d_2 u_2 \delta_2 \sum_{i=1}^{3} \frac{\partial u_i}{\partial p} \tag{1.43}$$

such that we get the differential equation

$$\frac{d}{dt}\left(\frac{\partial u_2(t,0)}{\partial p}\right) = -d_2[1+\delta_2 u_3(t,0)]\,\frac{\partial u_2(t,0)}{\partial p}\,. \tag{1.44}$$

This equation may, by using the abreviation $D_{it}=\exp(-d_i t)$ and

$$u_3(t,0) = u_3(0,0)\, D_{3t}/[1+\delta_3 u_3(0,0)\,(1-D_{3t})] \tag{1.45}$$

be integrated to

$$\frac{\partial u_2(t,0)}{\partial p} = \frac{\partial u_2(0,0)}{\partial p} D_{2t}/[1+\delta_3 u_3(0,0)(1-D_{3t})]^{(d_2\delta_2)/(d_3\delta_3)} \tag{1.46}$$

Using Eq. (1.46), with $(\partial u_2/\partial p)\,(0,0)=2P_2Bx$, and inserting Eq. (1.45) into the brace of Eq. (1.42) we get the condition for invasion with $D_i=\exp(-d_iT)$:

$$1<\frac{(P_2D_2)/(P_3D_3)}{[1+\delta_3 u_3(0,0)(1-D_3)]^{[(d_2\delta_2)/(d_3\delta_3)-1]}}\,. \tag{1.47}$$

Now, if we assume that the population size $x$ of the parental generation is at equilibrium corresponding to a population consisting of individuals entirely of genotype *aa*, i.e.,

$$x=(P_3D_3B-1)/[\delta_3P_3B(1-D_3)] \tag{1.48}$$

or

$$u_3(0,0)=(P_3D_3B-1)/[\delta_3(1-D_3)]\,, \tag{1.49}$$

then the bracket in the denominator of Eq. (1.47) simplifies to $P_3D_3B$ such that the condition for increase of $A$ when rare becomes

$$(P_2D_2B)^{1/\iota_2}>(P_3D_3B)^{1/\iota_3}\,. \tag{1.50}$$

We may generalize the above treatment to include differences in the birth factor $B$ among genotypes, and we may assume that the density dependent death rate of any specific genotype is influenced differently by the numbers of the other genotypes, i.e., is a function of the genotypic composition of the population. The death rate of genotype $i$ is then $d_i+\iota_{i1}u_1+\iota_{i2}u_2+\iota_{i3}u_3$, such that $\iota_{ij}$ is the death rate added on genotype $i$ for each individual of genotype $j$. This extension of the model thus allows differences between the genotypes in their vital requirements, and thereby differences in the competitive pressure between pairs of individuals. Taking these generalizations into account, the condition for increase of $A$ when rare becomes

$$(P_2D_2B_2)^{1/\iota_{23}}>(P_3D_3B_3)^{1/\iota_{33}}\,, \tag{1.51}$$

where the exponents only involve the density dependent death rates imposed on the population by the presence of the common homozygote (Poulsen, 1975).

Let us first confine our interest to the simple situation where the only difference between the genotypes is in the death rates, i.e., the relevant condition is Eq. (1.50) with $P_2=P_3=1$. Taking logarithms on both sides of Eq. (1.50) gives the equivalent condition

$$[\log(B)-d_2T]/\iota_2>[\log(B)-d_3T]/\iota_3\,. \tag{1.52}$$

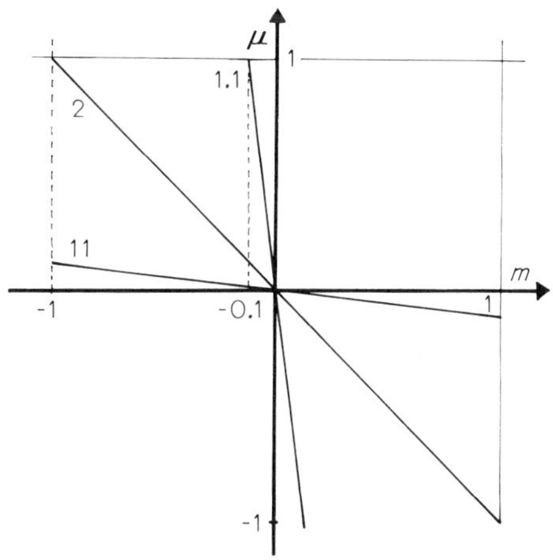


Fig. 5. The condition for increase of a new allele affecting the death rate parameters: $m=1-d_2/d_3$ and $\mu=1-\delta_2/\delta_3$. The curves shown are $m+(b/d_3-1)\,\mu=0$ for different values of $b/d_3$. The condition that $BD_3>1$ is that $m>1-b/d_3$ as shown for $b/d_3=2$ and $b/d_3=1.1$ by the broken lines

If we define the parameter $b=\log(B)/T$, which is the continuous time birthrate given a birth factor of $B$ in the time interval $T$, then Eq. (1.52) simplifies to

$$b/d_3-d_2/d_3>(b/d_3-1)\,(\iota_2/\iota_3) \tag{1.53}$$

showing that the condition for increase of $A$ is in terms of the relative changes in the death parameters but dependent on the ratio $b/d_3$. Designate the relative decrease in $d$ by $m$ and in $\iota$ by $\mu$, such that $d_2=(1-m)\,d_3$ and $\iota_2=(1-\mu)\,\iota_3$. Figure 5 shows the values of $m$ and $\mu$ for which $A$ increase for different values of $b/d_3$; $A$ increases for $(m, \mu)$ above or to the right of the lines shown in Figure 5:

$$m+[(b/d_3)-1]\,\mu>0\,. \tag{1.54}$$

It is evident that $A$ will increase whenever both $m$ and $\mu$ are positive because then the death rate of the heterozygote $Aa$ is lower than that of the homozygote $aa$ at all population densities. Outside this range, the new allele $A$ will have to decrease the basic death rate while sacrificing competitive ability (i.e., the density-dependent death rate) or vice versa. The parameter $b/d_3$ of the condition measures roughly the difference in population size between the newborn zygotes and the adults at equilibrium relative to a similar difference at low population densities, i.e.,

$$b/d_3=\lim_{u\to 0}\,[\log(\hat{x}/\hat{u})/\log(x/u)]\,. \tag{1.55}$$

Therefore, we may interpret high values of $b/d_3$ as meaning a highly competitive milieu, and low values as meaning a population only slightly influenced by competition. This interpretation is in accordance with the results as they emerge from Figure 5. For $b/d_3$ small (i.e., close to unity) even a very high improvement of the competitive ability can only pay off a very slight decrease in the basic death rate, whereas just a slight improvement in the basic death rate can pay off a considerable decrease in the competitive ability. On the other hand, for large values of $b/d_3$ very large increases in the basic death rate can be tolerated as long as this is associated by a decrease in $\iota$. So dependent on the initial situation in terms of $b/d_3$ the population will evolve towards higher competitive ability of the individuals or towards increasing its basic multiplication factor $BD$. However, except for very extreme cases, any population can evolve in both directions, but the likelihood of going in either direction may vary.

The equilibrium size of the population is likely to change if the new allele $A$ is increasing in the population. The initial change has not been determined explicitly. However, we may characterize the growth parameters of the genotypes by the corresponding carrying capacity, $K_3$ as given by Eq. (1.48) for the homozygote $aa$ and $K_2$ as the similar quantity in terms of the parameters of the heterozygote $Aa$ (this last "carrying capacity" is only to be considered as a formal definition because a population of pure heterozygotes is inconceivable in a sexual organism). For small $m$ and $\mu$ the ratio of the $K$'s may by Taylor expansion be approximated as

$$K_2/K_3 = 1+\mu+(\{(B-1)\,Td_3D_3/[(D_3B-1)(1-D_3)]\}-1)m$$
$$>1+\mu[1-(\{(B-1)\,Td_3D_3/[(D_3B-1)(1-D_3)]\}-1)\,[(b/d_3)-1]] \qquad (1.56)$$

where the evaluation comes from the condition for increase of $A$ given by Eq. (1.54). The approximation of the ratio of the $K$'s, and consequently the evaluation, depends only on the basic birth and death parameters of the common homozygote, $b$ and $d_3$, and numerical calculations of the factor of $\mu$ in the evaluation [Eq. (1.56)] show that it is always negative. The applied evaluation is the best possible in the sense that we may always choose $m$ and $\mu$ such that the two sides of the evaluation are arbitrarily close, with $A$ increasing in the population. Therefore, when $\mu$ is positive, the right hand side of the evaluation is less than unity, and it is possible to choose $m$ and $\mu$ such that $K_2 < K_3$ with $A$ increasing in the population. In this situation an allele that increases the competitive ability of its carriers may increase in the population even if its "carrying capacity" is lower than the actual population size.

This, however, does not prove that the population size initially declines, nor that the population will reach a new equilibrium where the population size is lower than the initial. Iteration of the model has, however, shown that the increase of a new allele in the population due to selection may in fact cause the population size to decline (Poulsen, pers. comm.).

In a more general treatment we have to turn to the condition [Eq. (1.51)] which by taking logarithms may be written as

$$r_2/r_3 > \iota_2/\iota_3\,, \qquad (1.57)$$

where $r_i$ is the continuous time growth rate corresponding to the multiplication factor $B_iP_iD_i = \exp(-r_iT)$. In this formulation of the condition [Eq. (1.57)] for the increase of the allele $A$ it becomes even more apparent that it is a comparison of the reproduction potential or unlimited growth rate with the competitive ability measured by the density dependent death rate per individual in the population. However, as is apparent from the definition, in this model the quantity $r$ is composed of the birthrate $b$, the death rate corresponding to the initial survival of the zygotes, $p=[\log(P)]/T$, and the death rate $d$, such that $r_i=b-p-d$. These three components are basically of a quite different nature, just to mention that $b$ corresponds to a discrete event whereas $d$ is a true continuous time death rate. Further, any sensible use of a model like this has to specify constraints on the parameters in terms of the physiology and environment of the organism that we have in mind. We will return to this point in a later section.

We will now turn to the model for discrete generation population growth [given by Eq. (1.10)] for studying density dependent selection. This model has proved useful in studying genetical variation in ecological models (Anderson, 1971; Charlesworth, 1971; Roughgarden, 1971, 1975). It may be viewed as an approximation to the discrete logistic model if the reproductive potential $R$ is small and positive (for another approximation, closely related to Eq. (1.19) see Clarke, 1972). However, the preceding results suggest cautious interpretation of the two parameters $K$ and $R$ in the linear model, as discussed in section 1.1.

Let $R_i$ and $K_i$ be the growth parameters of the three genotypes such that the recurrence equations between two generations are

$$\begin{aligned} x_1'-p^2x&=p^2xR_1(K_1-x)/K_1\,, \\ x_2'-2pqx&=2pqxR_2(K_2-x)/K_2\,, \qquad (1.58)\\ x_3'-q^2x&=q^2xR_3(K_3-x)/K_3\,. \end{aligned}$$

This formulation simplifies considerably by defining the fitnesses of the three genotypes as functions of $x$:

$$w_i(x)=1+R_i(K_i-x)/K_i\,, \qquad (1.59)$$

and observe that the total population size changes according to

$$x'=w_0(x)\,x \qquad (1.60)$$

where

$$w_0(x)=p^2w_1(x)+2pqw_2(x)+q^2w_3(x)\,. \qquad (1.61)$$

The gene frequency recurrence equation is now

$$p'=p(pw_1+qw_2)/w_0 \qquad (1.62)$$

where we for convenience have dropped the symbols showing that the fitnesses are functions of the total population size.

The equilibria of this model are found by solving the equations $p'=p=\hat{p}$ and $x'=x=\hat{x}$ which are obtained from Eq. (1.60) and Eq. (1.62). From Eq. (1.60) we have $w_0(\hat{x})=1$ and from Eqs. (1.59) and (1.61) we get

$$\hat{p}^2 R_1(K_1-\hat{x})/K_1+2\hat{p}\hat{q}R_2(K_2-\hat{x})/K_2+\hat{q}^2 R_3(K_3-\hat{x})/K_3=0\,. \tag{1.63}$$

Collecting terms in $\hat{x}$ gives the solution:

$$\hat{x}=\frac{\hat{p}^2 R_1+2\hat{p}\hat{q}R_2+\hat{q}^2 R_3}{\hat{p}^2 R_1/K_1+2\hat{p}\hat{q}R_2/K_2+\hat{q}^2 R_3/K_3}\,. \tag{1.64}$$

Thus, the equilibrium population size is the weighted harmonic mean of the carrying capacities $K_i$ with the weights $R_i$ (Anderson, 1971; Charlesworth, 1971). Using this and the equlibrium equation obtained from Eq. (1.62) which is

$$\hat{p}[\hat{p}R_1(K_1-\hat{x})/K_1+\hat{q}R_2(K_2-\hat{x})/K_2]=0 \tag{1.65}$$

then we get an equation with which we could determine the equilibrium gene frequency $\hat{p}$, and thereby through Eq. (1.64) the whole equilibrium structure. However, this is a very complicated equation and we will not follow a general solution here; it is presented by Anderson (1971). We can, however, state the conditions for a solution. If we substract Eq. (1.65) from Eq. (1.63) we get

$$\hat{q}[\hat{p}R_2(K_2-\hat{x})/K_2+\hat{q}R_3(K_3-\hat{x})/K_3]=0 \tag{1.66}$$

which is symmetric to Eq. (1.65). From Eq. (1.65) and (1.66) it is seen that $\hat{p}=0$ and $\hat{x}=K_3$ and $\hat{q}=0$ and $\hat{x}=K_1$ are trivial solutions to the equations. Solutions with nonzero values of $\hat{p}$ and of $\hat{q}$ may be found from the linear equations which arise from Eqs. (1.65) and (1.66) by taking away a factor $\hat{p}$ and $\hat{q}$ respectively. Since $q=1-p$, we get a solution for $\hat{p}$ in both equations and by equating them we get an equation for $\hat{x}$. However, $\hat{p}$ being a frequency we must require that it is positive and less than 1. The solution we get from Eq. (1.65):

$$\hat{p}=[R_2(K_2-\hat{x})/K_2]/[R_2(K_2-\hat{x})/K_2-R_1(K_1-\hat{x})/K_1]\,, \tag{1.67}$$

is positive and less than unity only if $K_1<\hat{x}<K_2$ or $K_2<\hat{x}<K_1$. By a symmetric argument with respect to the solution from Eq. (1.66) we see that a valid solution is characterized by one of the two following conditions: (1) if $\hat{x}<K_2$, then $K_1$ and $K_3$ are less than $\hat{x}$, or (2) if $\hat{x}>K_2$, then $K_1$ and $K_3$ are greater than $\hat{x}$. Thus the system has only a valid nontrivial equilibrium if either $K_2$ is greater or smaller than both $K_1$ and $K_3$, i.e., either over- or underdominance is required with respect to the carrying capacities. If this is the case, a few calculations show that an internal equilibrium indeed exists.

By an argument quite similar to the one leading to Eq. (1.34) we can write

$$p'-p=[pq/(2w_0)]\partial w_0/\partial p\,, \tag{1.68}$$

and thus the change in gene frequency is such that $w_0(x)$ locally increases as a function of $p$. However, note that when $p$ changes *and* $x$ changes then the fitnesses change so we cannot conclude from Eq. (1.68) that $w_0$ increases. In fact it does not because for small populations it is of the order $1+R$ whereas at equilibrium it is 1.

Another maximum principle, originally suggested by MacArthur (1962) for a continuous generation model and translated into the present model by Anderson (1971) is the following. Define the functions

$$R(p)=p^2R_1+2pqR_2+q^2R_3\,,$$
$$K(p)=R(p)/[p^2R_1/K_1+2pqR_2/K_2+q^2R_3/K_3] \tag{1.69}$$

and observe that from Eq.(1.64) we have that $K(\hat{p})=\hat{x}$. The interpretation of the functions $K$ and $R$ is straightforward; they are the mean carrying capacity and the mean reproductive potential respectively. If the selection forces are small compared to the rate at which the population size goes to equilibrium, then we may expect that the population size at any instant is given by $K(p)$. To see the relevance of this, observe from the definition [Eq.(1.61)] of $w_0$ that

$$w_0=1+R(p)-[R(p)/K(p)]x \tag{1.70}$$

which differentiated with respect to $p$ yields

$$\partial w_0/\partial p=(1-x/K)\partial R/\partial p+(x/K)(R/K)\partial K/\partial p\,. \tag{1.71}$$

When $x\approx K(p)$ then Eq. (1.68) and Eq. (1.71) yield the approximative relation

$$p'-p=[pq/(2w_0)]\,[R(p)/K(p)]\,\partial K(p)/\partial p\,. \tag{1.72}$$

Thus the direction of increase in $p$ (when $x\approx K(p)$) is to increase $K(p)$. [From Eq.(1.72) it is also seen that the internal equilibrium point is stable for overdominance with respect to the $K$'s and unstable for underdominance.]

The procedure for reaching this result is rather sloppy; the formal arguments given above are outlined by Anderson (1971) and Roughgarden (1975). (Roughgarden also generalizes the principle to include the possibility of more species.)

For the more simple situation, where $R_i=R$, and the genotypes only differ with respect to the carrying capacities, then the question becomes more simple and the internal equilibrium is of the form [Eq.(1.32)] with the $K$'s replacing the $P$'s. Further, Eq.(1.72) is more reliable because $R$ is a constant and from Eq.(1.71) and Eq.(1.68) we get the exact relation

$$p'-p=[pq/(2w_0)]\,(Rx/K^2)\partial K(p)/\partial p \tag{1.73}$$

which shows that the gene frequency will here always change in a direction where $K(p)$ increases.

These results, i.e., that the carrying capacity is always maximized at equilibrium is in contrast to our previous results with the Poulsen model. This is due to the basic differences in the parameterization and formulation of the two models. In the Anderson-Roughgarden model the equilibrium population size $K$ of a homozygous population is considered to be directly influenced by the genetic variation, and superiority in the competition at high population sizes is only allowed as an increase in the carrying capacity. Thus, almost by definition, the increase of a superior genotype will lead to an increased population size. The formulation used in the Poulsen model is fundamentally different. Here the equilibrium population size is found as a balance between the density dependent death and birthrates, and the total population size is only influenced by genetic variation through the evolution of the birth, death and competition parameters of the individuals.

The discrepancy of the two models is therefore expected, but both being simple and somewhat artificial models, it is still a question as to which gives the closest prediction in a natural population. Superficially, the much more detailed Poulsen model seems preferable, and we certainly believe it to be so in many cases. However, the Anderson-Roughgarden model may well in many cases summarize complicated interactions between the parameters involved in birth, death and competition, where the Poulsen model would be too complicated to answer even the simplest questions. This virtue of mathematical simplicity of the Anderson-Roughgarden model will become handy in the later discussion of evolution of species interactions (Chapt. 3), but in the following exposition of life cycle characteristics of single species populations, the Poulsen model is the natural reference.

## 1.3 Competition and Bacterial Growth in Chemostats

To illustrate the concepts of reproduction potential, carrying capacity, and competition, let us explore the simplest possible population model conceivable, that of the growth of bacteria in a chemostat. The chemostat is used for growing microorganisms in a physiological steady-state under constant environmental conditions. The theory of this system was developed independently by Monod (1950) and by Novick and Szilard (1950). The chemostat is the best understood, albeit very unnatural ecological system.

A chemostat (Fig. 6) consists of a culture vessel with an overflow, and with the addition of a sterile substrate at a constant rate, $f$. The *dilution* or *washout rate* of the system is given as $D=f/V$ where $V$ is the volume of the culture vessel. The cells growing in the culture are assumed to be distributed homogeneously in the volume so that their washout rate is also $D$; this is usually accomplished by a magnetic stirrer. If we add a bacterial inoculate to the chemostat and we set $D<\mu_m$, where $\mu_m$ is the unlimited growth rate of the bacterial strain with the particular substrate used, then the bacterial population will increase according to

$$dx/dt = x[\mu(x) - D], \qquad (1.74)$$

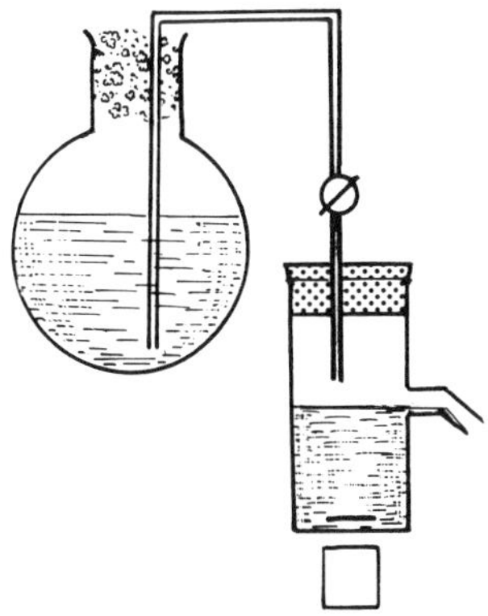

Fig. 6. A chemostat, schematically drawn

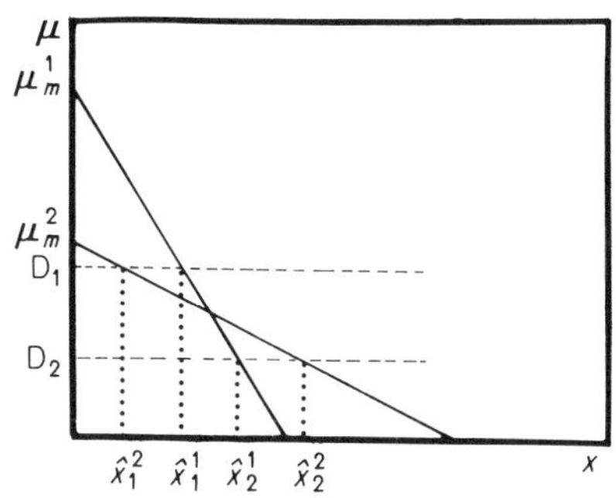


Fig. 7. The relation between specific growth rate, $\mu$, and population density, $x$, in a chemostat for two strains of bacteria and their equilibrium population sizes for two different dilution rates, $D_1$ and $D_2$. Strain 1 will win at the high dilution rate $D_1$ but will be excluded at $D_2$ when the two strains are grown in the same chemostat

where $\mu(x)$ is the density dependent growth rate. As $x$ increases, $\mu(x)$ will decrease due to the competition for the substrate between the bacteria. This decrease will continue until $\mu = D$, and we will arrive at a steady-state situation with a population density $\hat{x}$, the carrying capacity, where the cells divide at a constant rate $D$. Figure 7 shows two examples of the relationship between $\hat{x}$ and $D$ for the case of a linear functional form of $\mu(x)$. One (strain 1) with a high value of $\mu_m$ and a low ability for competition, i.e., a rapid decrease of $\mu(x)$ with $x$, and the other (strain 2) with a low value of $\mu_m$ but a high competitive ability, i.e. a slow decrease of $\mu(x)$ with $x$. It can be seen that strain 1 can sustain a higher population size than strain 2 at high dilution rates, whereas the opposite is the case at low dilution rates.

We can imagine the two strains of bacteria in Figure 7 as two genotypes of the same species, and we can then from the figure predict the outcome of the growth of the two types in a mixed culture for different dilution rates. If the two strains are using the same substrate qualities at the same rate per unit population size, then the growth rate of both strains is a function of the total population size in the chemostat. At the dilution rate $D_1$ both strains will grow as long as the total population size is less than $\hat{x}_1^2$. Reaching this population size strain 2 will cease to grow, whereas strain 1 will continue growing, such that the total population size will grow and thereby cause a total population size where the population of strain 2 will start declining. Thus, at dilution rate $D_1$ strain 1 will be superior in the competition with strain 2, and we will end up with a monoculture of strain 1. At dilution rate $D_2$ the opposite situation will emerge, and we will end up with a monoculture of strain 2. Note that in any case, the strain with the highest $\hat{x}$, the carrying capacity, will win.

In order to discuss the significance of this in some more detail, we will consider bacteria which are limited by their energy (carbon) source. It has been shown empirically that bacterial growth as a function of a limiting, dissolved carbon source can be described by

$$\mu = \mu_m s/(K_s + s) \tag{1.75}$$

where $s$ is the concentration of the limiting substrate and $K_s$ is a constant measuring the substrate concentration corresponding to the growth rate $\mu_m/2$. [Eq. (1.75) has the form of the Michaelis-Menten equation describing enzyme kinetics.]

Equations (1.74) and (1.75) provide a complete description of the kinetics of the chemostat. Combining the two equations we have

$$dx/dt = x[\mu_m s/(K_s + s) - D]$$

which provides an expression for the substrate concentration at steady state, viz.,

$$\hat{s} = DK_s/(\mu_m - D)\,. \tag{1.76}$$

Thus the steady state substrate concentration decreases with decreasing dilution rate in accordance with the fact that a low dilution rate gives a low washout rate of the bacteria but a higher steady state population and a more intense competition.

We may take small values of $K_s$ to mean high "substrate affinity", in the sense that the growth rate [Eq. (1.75)] increases rapidly to $\mu_m$ at a relatively low substrate concentration:

$$\mu_m - \mu = \mu_m/(1 + s/K_s)\,. \tag{1.77}$$

Conversely, high $K_s$ values mean that $\mu$ does not approach $\mu_m$ until the substrate concentration is high. In naturally occurring strains of bacteria, however, the two parameters of the growth rate seem to be highly correlated. A strain having a small $K_s$ and thus a high competitive ability with respect to a limiting resource, trades this off by having a low $\mu_m$, and a strain with a high $\mu_m$ will have a high $K_s$, too. As seen from Figure 8 we can expect bacteria with high $\mu_m$ and $K_s$ values to win in chemostats with high dilution rates, or equivalently, high

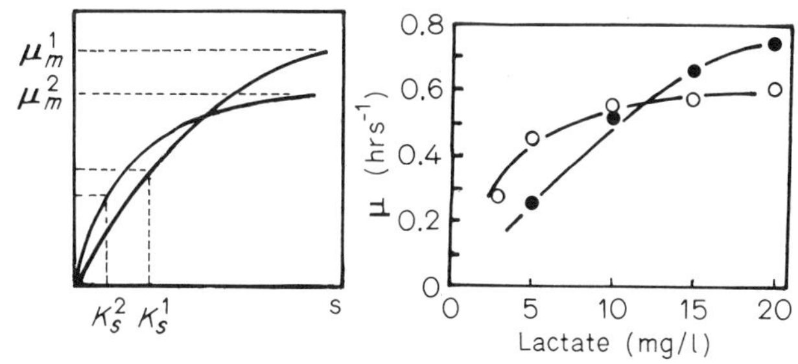


Fig. 8. *Left:* The specific growth rate of two strains of bacteria as function of the concentration of a limiting substrate. *Right:* The growth rates of a *Spirillum* sp. (*open circles*) and of a *Pseudomonas* sp. (*filled circles*) in lactate limited chemostats. (After Veldkamp and Jannasch, 1972)

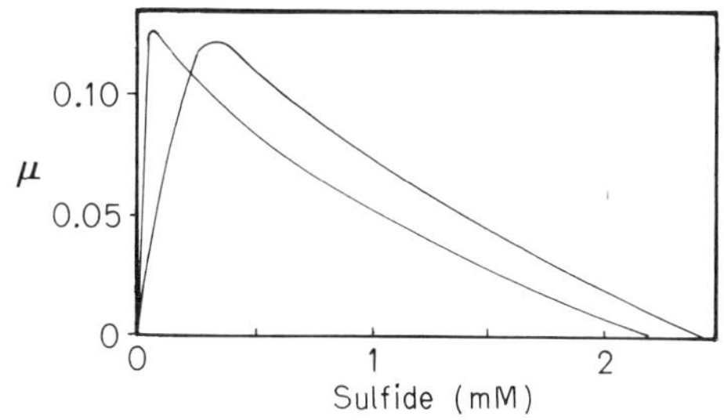


Fig. 9. The growth rates of two photosynthetic sulfurbacteria as functions of sulfide concentration. (After Veldkamp and Jannasch, 1972)

substrate concentrations in the reservoir. On the other hand, bacteria with low $\mu_m$ and $K_s$ values are expected to win at low dilution rates or low substrate concentrations. These two types of bacteria are in fact found in nature (Fig. 8). They were already recognized by the early microbial ecologists as the *zymogenous* microflora which utilizes high concentrations of organic substrates (such as will occur periodically, e.g., close to fresh carrion) and the *autochonous* microflora which generally dominates undisturbed soils which have only low concentrations of easily decomposable, dissolved organics (see Winogradsky, 1949). The two extreme types (large $\mu_m$ and large $K_s$ vs. small $\mu_m$ and small $K_s$) also correspond to the *r*- and the *K*-selected species of MacArthur (1962). In a chemostat with a given value of *D* and of the substrate concentration in the reservoir, only one strain (or genotype, or species) will persist, but as mentioned, a chemostat is not a natural system. In nature, the temporal and spatial heterogeneities make the persistence of both types intelligible (see also Chap. 4).

Veldkamp and Jannasch (1972) in a review on competition in the chemostat discuss an example with two photosynthetic sulfurbacteria growing in sulfide limited chemostats. They found two forms of which one had the lowest value of $\mu_m$ as well as the highest value of $K_s$. To the extent that sulfurbacteria always are sulfide limited in nature, this species would not exist. However, it could be shown that at very high sulfide concentrations, when the substrate becomes toxic and limits both forms, then the form which is inferior as competitor at all optimal and suboptimal concentrations becomes the superior form (Fig. 9). This is an interesting example, because the species has "traded" a high value of $\mu_m$, not for an improved utilization of a scarce resource in a competitive situation, but for coping with a resource when it becomes toxic at high concentrations. This is a situation that photosynthetic sulfurbacteria would experience early a succession.

## 1.4 Reproductive Effort of Populations

In order to study the effect of selection with respect to features which increase the unlimited growth rate vs features which increase competitive ability, one could compare the relative amount of resources allocated to reproduction in different populations. This is often difficult in practice. In flowering plants, however, the reproductive effort can be estimated by the weight of the flowers compared to the weight of the vegetative organs.

Such a study was carried out by Gadgil and Solbrig (1972) on populations of dandelions (*Taraxacum officinale*). This plant has the advantage of being parthenogenetic and fertilization never takes place. Gadgil and Solbrig (1972) studied three localities situated within a distance of 1/2 km, which differed with respect to environmental conditions. The first one was a dry, frequently mowed lawn with many bare patches due to disturbance. The dandelions in this locality were probably exposed to a considerable density independent mortality. A second locality was somewhat more protected and the third one had a nearly totally undisturbed vegetation and was much more moist; here the dandelions could be expected to be limited mainly by competition.

Gadgil and Solbrig (1972) isolated four biotypes of dandelions from these three localities. The types were characterized by isozyme patterns and by morphological characters which were retained when the plants were grown under identical and controlled conditions. There were considerable differences between the biotypes with respect to numbers of flowering heads, the amount of nonreproductive tissue produced, and in the first time of flowering. The biotype allocating most energy into flower production was predominantly found in the first mentioned dry and disturbed locality, and the form producing more leaves and fewer flowers was found practically exclusively in the moist, undisturbed locality. The two intermediate types were mainly found in the intermediate locality. Further, the biotype found in the disturbed habitat produced flowers earlier than the rest of the biotypes.

This study represents one of the few examples of the effect of density dependent selection on reproductive effort which are both convincing and amenable to analysis.

The variation in clutch size of different birds species and of the same species of birds in different localities has attracted considerable interest in the literature.

Clearly birds do not in general lay the maximum number of eggs which they are physiologically capable of producing since most birds will readily replace eggs which have been removed from their nest. Rather clutch size expresses a compromise that maximizes the fitness of the parent birds. An increase in clutch size would lead to an insufficient feeding of the youngs and, therefore, an

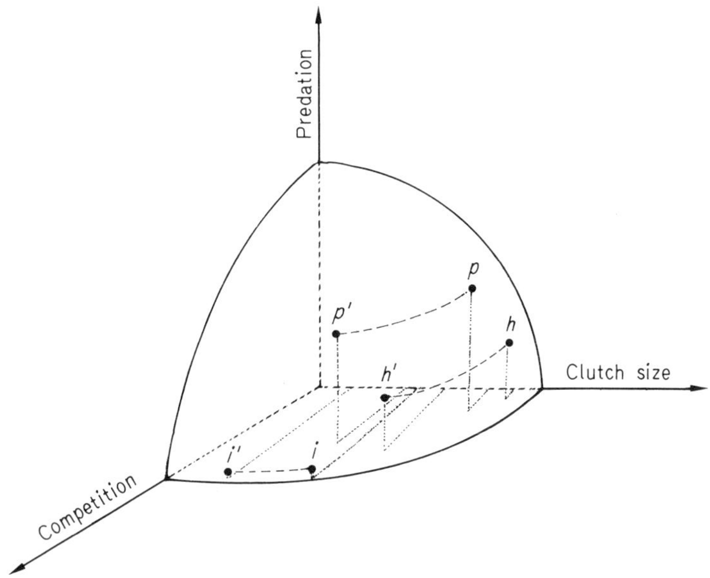


Fig. 10. The changes in the allocation of resources between clutch size, avoiding predators, and competitive ability for three groups of birds (*p*: passerines; *h*: hole nesters; *i*: insular birds) in temperate climates and in the tropics ('). (Modified after Cody, 1966)

excess mortality, whereas a decrease in the clutch size would not be compensated sufficiently by a decreased mortality (Lack, 1947). There is, however, a wide variation in clutch size between different kinds of birds. For example, the California condor first matures after five years and then produces only one egg every second year (Mertz, 1971), whereas ducks may have clutch sizes of about ten.

Cody (1966) discusses clutch sizes of birds in terms of a graphical model shown in Figure 10. It assumes that birds may allocate their resources (represented by the spherical surface in the figure) between clutch size, competitive ability and predation protection (especially nest predation). As competitive interactions, for examplc, bccome more intense, selection will favor fewer resources being allocated to egg production and more resources to competitive ability, e.g., to maintain a territory. If density independent mortality (unpredictability of food availability or an unpredictable climate) is more important, then more resources will be allocated to increase clutch size (however, see Mountford, 1973). Nest predation will, according to Cody (1966) tend to decrease clutch size since the parents then will have to spend more time protecting the nest and thus less time can be spent gathering food. This should explain the high clutch sizes characteristic for hole nesters.

Cody's model does predict some of the characteristic trends in clutch size, e.g., the lower clutch size found in nearly all tropical birds compared to their temperate relatives (Fig. 11). This is explained by the higher predictability of environmental factors in lower latitudes, which will increase the intensity of competition (Chap. 3). Cody also mentions that nest predation is more frequent

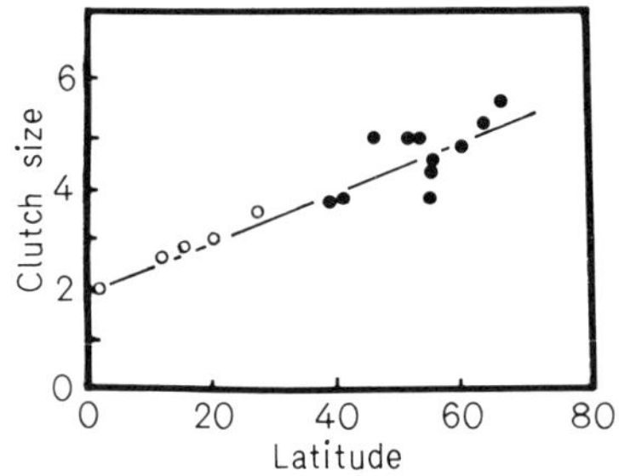


Fig. 11. The average clutch size of buntings (*Emberiza* spp.) at different latitudes. *Open circles*: southern hemisphere; *filled circles*: northern hemisphere. (After Cody, 1966)

in the tropics. The low clutch sizes of oceanic birds (the albatross, for example) is similarly explained by the constant supply of food resources and the nearly constant climatic conditions of oceanic islands.

The model of Cody is in many respects of a so general a nature that it may indeed explain anything found regarding clutch sizes. It is, for instance, not trivial that decreased predation will lead to an increased clutch size. This is certainly not the case for birds breeding on oceanic islands; here the model simply claims that in say albatross colonies, competition is even more intense so as to overshadow the effect of absent predation. If hole nesters are relieved from mortality due to predation as compared with their relatives nesting in the open, why then are the hole nesters not subject to a more intense competition which tends to decrease clutch size? The model is useful in systematizing the observations on clutch sizes of birds and give some predictions which may be tested but it is too general to contribute to a deeper insight in density dependent selection.

This theory describes the variations in the mean clutch size of different bird populations. However, many bird species show considerable variation in clutch size within populations. This may be understood as an effect of varying environments (Mountford, 1973) along the same lines as discussed in the next section.

### 1.4.1 Delayed Germination of Annual Plants

It is well known that some desert plants show delayed germination, i.e., only a certain fraction of seeds from a given year will germinate the following spring; of the remaining, a fraction will germinate in the following spring, and so on. This is interpreted as an adaptation to an unpredictable climate; some years will be too dry to allow the plants to produce new seeds. Consequently, an annual plant is believed to increase its reproductive success by producing seeds with delayed germination; this will increase the probability that at least a fraction of the seeds will germinate in a favorable year. Cohen (1967) proposed a model describing this phenomenon. Since this model is also discussed in detail by MacArthur (1972) we will only treat it briefly here. In the present context the model is interesting because it shows that an unpredictable climate and the resulting density independent mortality may lead to adaptations which decrease the short term reproduction potential by maximizing the long-term growth rate.

We start with $N$ seeds one winter and suppose that every spring a fraction $G$ germinates. In a good year, the $NG$ seeds produce each $S$ seeds, whereas in bad years the $NG$ plants fail to reproduce. In both kinds of years $(1-G)N$ will remain in the soil. After a good year, there will therefore be $SNG+N(1-G)=N(SG+1-G)$ seeds in the soil, and after a bad year only $(1-G)N$ seeds. If the fraction of good and bad years are $p$ and $1-p$ respectively, then in a sequence of $T$ years, where $T_1$ were good and $T_2$ were bad, the original $N$ seeds will have multiplied by a factor

$$R^T=(GS+1-G)^{T_1}(1-G)^{T_2}.$$

In the long run, $T_1$ will be close to $pT$ and $T_2$ to $(1-p)T$ such that in the average the multiplication of the population of seeds per year is

$$R=(GS+1-G)^p(1-G)^{(1-p)}. \tag{1.78}$$

This multiplication factor is maximized for a germination frequency $G$ where

$$dR/dG=(S-1)p(GS+1-G)^{(p-1)}(1-G)^{(1-p)}-(1-p)(GS+1-G)^p(1-G)^{-p}=0,$$

or

$$G=(Sp-1)/(S-1) \tag{1.79}$$

which is close to $p$ for large $S$.

Thus in order to maximize the reproductive success a plant should produce seeds of which approximately a fraction corresponding to the frequency of good years should germinate each year.

### 1.4.2 Demographic and Energetic Components of the Unlimited Growth Rate

When we leave comparisons of related species or genotypes of the same species for which measures such as say the division rate of bacteria or the number of flower heads of dandelions are reasonable measure of the reproductive effort and start to compare organisms covering a wider taxonomical or ecological range, then it is nessesary to arrive at a better understanding of the components of the reproductive potential since it is these components which are directly affected by selection.

The exponential growth rate in an unlimiting habitat is always a function of two kinds of events, i.e., the birth and the death of single individuals, and as previously discussed in this chapter, we may in the simplest situation use the model that the birth and death events are independent such that $r=b-d$, where $b$ and $d$ are the instantaneous birth and death rates respectively. However, if we allow an extended life span of the individuals with the death rate and the fecundity of individuals dependent on age, then the birth and death of individuals are no longer independent, but both are dependent on the age structure of the population. Consider a *cohort* of animals (a collection of individuals born at the same time) from birth at time 0 to the maximum life span $\omega$. Designate by $l_a$ the survivorship function at age $a$, i.e., the proportion of the cohort that survives from birth to age $a$, and designate by $m_a$ the fecundity of individuals of age $a$, i.e., the number of off-spring produced by individuals of age $a$. Each individual of the cohort on the average at age $a$ then contributes $l_a m_a$ individuals to the next generation, i.e. to the cohort born at time $a$. The whole reproductive contribution of the cohort born at time 0, i.e., the net reproduction per individual per generation, is

$$R_0 = \int_{a=0}^{a=\omega} l_a m_a da\,. \tag{1.80}$$

If the survivorship function and the fecundity function are unchanged through time, then the population will after some time grow at an exponential rate $r$ and attain a stable age structure characteristic for the value of $r$ and the function $l_a$ (for a formal proof see Pielou, 1969; Williamson, 1972). From the equation for exponential growth: $x(t) = x_0 \exp(rt)$ [Eq. (1.2)], we have for a time span of one generation, $T$, that

$$x_{t+T}/x_t = R_0 = \exp(rT)$$

or

$$r = (\log R_0)/T\,. \tag{1.81}$$

For a population to grow at the rate $r$ with a stable age structure we must require that each age class is growing at the same rate $r$. Now consider such a population which at time $t$ gives birth to $y$ individuals. At time $t-1$ there were, therefore, $y\exp(-r)$ newborn individuals, at time $t-2$ there were $y\exp(-2r)$, and at time $t-a$ there were $y\exp(-ar)$ individuals of age 0. At time $t$, the total population size including all age classes is

$$\int_{a=0}^{a=\omega} y l_a \exp(-ra) da$$

individuals which give birth to

$$\int_{a=0}^{a=\omega} y l_a m_a \exp(-ra) da = y \tag{1.82}$$

individuals at time $t$. Equation (1.82) is an equation connecting the demographic parameters $l_a$ and $m_a$ to $r$ (by taking the factor $y$ away on both sides of the equation). From Eq. (1.81) we may then get the generation time $T$.

When the adaptive significance of different reproductive patterns are studied, the two parameters, viz., the survivorship function and the age specific fecundity must be discussed separately since natural selection affects them directly. However, we will in the following restrict our attention to the growth rate $r$ to find general characterizations of its properties through a diversity of organisms.

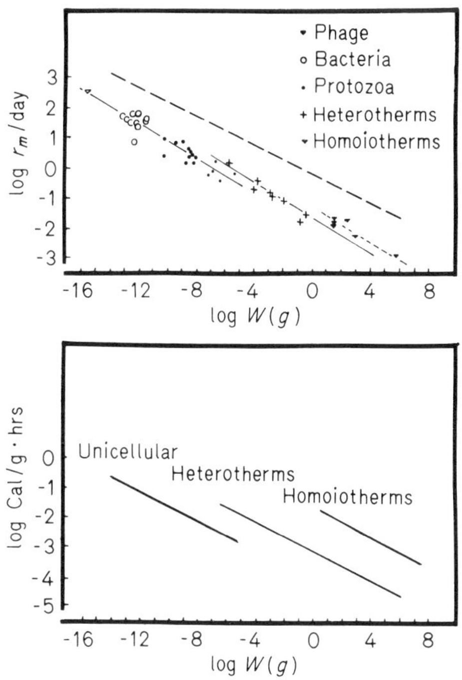


Fig. 12. The maximum unlimited growth rates for a number of organisms as functions of body weight and (*below*) the relationship between metabolic rate per unit weight and body weight for the animal kingdom. The broken line in the upper figure shows the slope of the metabolic rate/weight relationship. (After Fenchel, 1974)

Let us now look at the energetic significance of the unlimited growth rate. Clearly $r$ may be taken as a measure of the production of biomass of a population which grows exponentially with the rate $r$. For any species the value of the unlimited growth rate is dependent on the environmental factors. However, for each species there will be at least one set of environmental factors which yields a maximum growth rate, $r_m$, and this value may be considered a species characteristic. Such environmental conditions which yield the maximum growth rate can often be approximated closely in laboratory cultures. Fenchel (1974) collected data on the maximum growth rates for different species and related these values to body size (Fig. 12). The relation can be described by the equation

$$r_m = aW^n \tag{1.83}$$

where $W$ is the weight, $n$ is a constant taking the value $-0.275$, and $a$ is a constant taking three different values for unicellular organisms, and for heterotherm and homoiotherm metazoans respectively. This relationship is reminiscent of a well-known relationship between body size and metabolic rate per unit weight (Fig. 12). It is intelligible that the ratio between metabolic rate per unit weight and the productivity of populations can only vary within certain limits so that a correlation between these two parameters could be expected. With respect to the graphs we should note the following.

1. The curves probably reflect some upper limit of $r_m$ which is energetically possible for each size group within each of the three types of organisms. This is so because practically all of the species shown are either species frequently used in laboratories, domestic animals or pests, all of which are characterized by high potential growth rates. Species within each size group which do not have these characteristics, as for instance the autochonous microflora or other species which are found in climax communities, would yield points below the lines [examples are man and the California condor; according to Mertz (1971) the latter has a $r_m$ of only about 0.0003/day which is much lower than would be expected from a homoitherm animal of that size according to Fig. 12].

2. During the evolution from unicellular organisms to metazoa and from heterotherms to homoiotherms an increase in potential growth rate took place but due to the increased structural complexity, energetic efficiency decreased. Thus a homoiotherm mammal has a $r_m$ which is on the average 2.3 times higher that that of a reptile of similar size, but the mammal has a metabolic rate which is 55 times higher, mainly due to the higher heat production. The increased reproduction potential of birds and mammals relative to that of reptiles has mainly been achieved by a shorter development time (Fenchel, 1974). The smaller slope of the $r_m$—body weight relationship ($-0.275$) relative to that of the metabolic rate—body size relationship ($-0.249$) also shows that with increasing size and thus with increasing structural complexity an increasing amount of energy resources of species are allocated for maintenance.

## 1.5 An Evolutionary Dichotomy

The previous sections provide examples of the physiological constraints on rates of growth and production of eggs on the one hand, and on the ability of sustaining life in competitive situations or in other situations of stress on the other hand. The evolution of the life cycle and reproduction characteristics of a species must occur within these constraints. We have seen that within the physiological constraints of bacteria, we find strains with high unlimited growth rate and poor competitive ability and other strains with high competitive ability and low unlimited growth rate. In dandelions the same variation of types occurs, and similar phenomena are observed in bird species. This dichotomy in the life and reproduction of organisms is sometimes referred to as $r$- versus $K$-strategy. The evolutionary forces producing them are named $r$- and $K$-selection (MacArthur, 1962), referring to selection taking place during the initial phase of growth in the logistic model, or to selection in a population at its equilibrium size, $K$.

In populations mainly controlled by density independent mortality with large fluctuations in the population sizes this initial phase of growth is recurrent, and selection will favor features that tend to increase $r_m$, such as increased litter sizes or shortened development time (within the physiological constraints discussed above). Under other circumstances where populations are mainly controlled by inter- and intraspecific competition and are characterized by constant population sizes, selection will mainly favor adaptations which increase competitive ability which at the same time may mean a sacrifice of say litter size. However, our previous discussion show that the argument on $r$- versus $K$-selection is only meaningful when closely related species are compared, unless, of course, one wants to express the triviality that selection does not exclusively favor a maximized $r_m$ (in which case we would all be bacteria). But to claim that, e.g., butterflies are more $r$-selected than deer (Pianka, 1974), is to render the whole concept meaningless.

Essentially, the concept of $r$- and $K$-selection is that individuals of a population limited mainly by competition will increase their competitive ability, while in populations which are kept far below the carrying capacity by density independent mortality selection will favor features which increase the unlimited growth rate. However, selection will always favor features which give small positive deviations from the *realized* value of the growth rate, which for any population averages *zero* over any extended period of time.

The evolutionary dichotomy indicated from the observations of closely related species may be compared to our theoretical results of the Poulsen model. First, let us consider the situation where the new allele in the population, $A$, does not influence the competitive ability of its carriers but changes the birth factor $B$ and the basic death rate $d$. This is thus a case of pure $r$-selection (Hairston et al., 1970; Pianka, 1972). These two parameters must, however, be specified within the physiological constraints of the organism we have in mind. A simple example is that the organism has a certain amount of energy to spend for the purpose of reproduction, $E$, such that the birth factor is $B=E/s$, where $s$ is the size of an offspring. The model may then specify that the basic death rate decreases with the size of the offspring, i.e., $d=d_0+f(s)$ where $d_0$ is a constant and $f(s)$ is a decreasing function (Smith and Fretwell, 1974). Let us specify the function $f$ as a member of the family of functions $f(s)=e/(s^\kappa+k)$ where $e$, $\kappa$, and $k$ are positive constants. Two distinct possibilities of evolution exist for these models. For $e\kappa<4k$ selection will favor a decrease in offspring size for every initial size, i.e., the birth factor will increase. On the other hand, for $e\kappa>4k$ there exists an offspring size which is optimal, and to which the population will converge in the course of evolution, given the initial offspring size is above a certain suboptimal level. Below this level, selection will favor an ever declining offspring size. Examples of both these situations are shown in Figure 13 for different shapes of the function $f$ (all with the normalisation of $T=1$).

The analysis of the above model proceeds as follows: The condition for increase of $A$ is from Eq. (1.51) and with $D=\exp(-d)$:

$$(E/s_2)\exp\{-[d_0+f(s_2)]\}>(E/s_3)\exp\{-[d_0+f(s_3)]\}, \tag{1.84}$$

i.e., the evolution is determined by the variation of the function $(1/s)\exp[-f(s)]$. Maxima and minima of this function are determined by equating the derivative to zero, which produces the equation

$$\partial f(s)/\partial s=-1/s \tag{1.85}$$

or

$$(s^\kappa)^2-(e\kappa-2k)s^\kappa+k^2=0 \tag{1.86}$$

which has two positive solutions when $e\kappa>4k$. By consideration of the second derivative of the function in Eq. (1.84) it may be shown that the larger root of Eq. (1.86) always corresponds to a maximum and that the smaller root corresponds to a minimum. Thus below the minimum the function $(1/s)\exp[-f(s)]$ is increased by going towards zero, and above the minimum it is increased by going towards the larger root of Eq. (1.86).

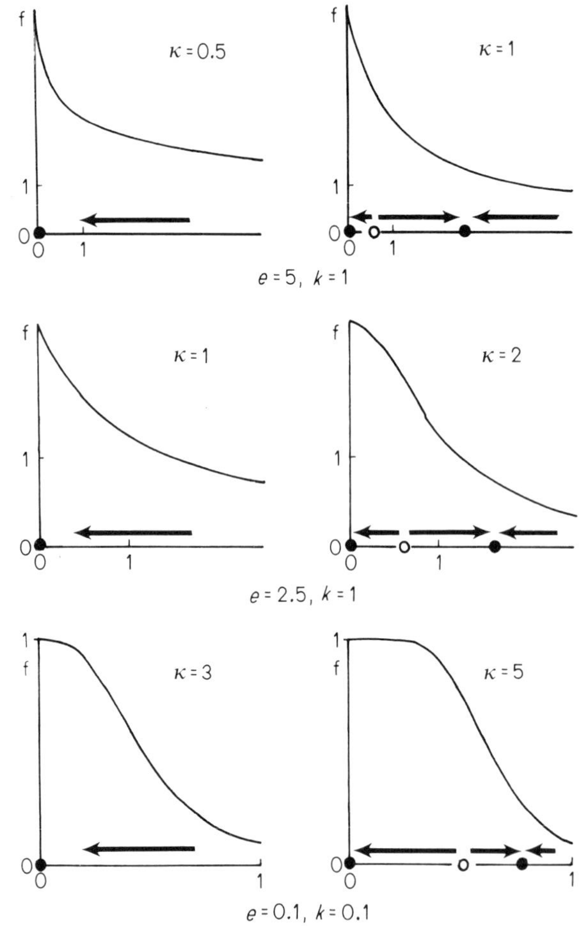


Fig. 13. Examples of the functional form of the size-dependent death rate. The figures on the left correspond to the situation where decreased offspring size is always favored, and those on the right correspond to the situation where an optimal size exists. (●, optimal point. ○, point separating the directions of evolution)

These considerations are founded on the rather arbitrary choice of the form of the function $f$. However, the above conclusion that in some cases there is an optimal offspring size to which the population may converge is valid whenever the function $f$ at some point increases sufficiently with decreasing size, e.g., if $k=0$ the optimal offspring size corresponds to a global maximum in fitness. The conclusion that the evolution of offspring size may depend on the initial size is founded on the assumption that the death rate does not increase without bounds at zero. However, the offspring size of zero, or in practice any sufficiently low size, is inconceivable for any real organism. But this lower bound may then be considered as the optimal offspring size for populations having an initially low offspring size.

The evolution of offspring number or offspring size through $r$-selection may thus lead to the attainment of an optimal evolutionary stable situation with a limited number of large offspring in some situations, and to an ever increasing number of tiny offspring in other situations. Thus from this analysis it is conceivable that both sharks and sunfish *(Mola mola)* developed their respective reproductive patterns by $r$-selection; the former produce few ($<10$) large eggs or liveborn fish yearly while the latter produce more than $10^8$ eggs.

In the more general case, where the new allele is affecting both the reproductive potential and the competitive ability, a similar dichotomy is indicated by the analysis on p. 17. In any situation an allele which increases the unlimited growth rate by sacrificing competitive ability may increase, and so may an allele which increases the competitive ability at the expense of the unlimited growth rate (both under the constraints shown in Fig. 5). In any particular case the evolutionary change in either direction must be within the physiological constraints allowed by the organism, so the model will predict that organisms may evolve either mainly to increase the unlimited growth rate ($r$-selection) or mainly to improve competitive ability ($K$-selection).

Another indication from the theoretical investigation of the Poulsen model is that $K$-selection, i.e., selection for increased competitive ability, does not necessarily result in an increased $K$. This is in contrast to the clear results from experiments with bacteria in chemostats. However, other kinds of competitive interactions are conceivable in higher organisms. In the bacteria, the competition does not involve any direct interaction between the individuals in the population, it is merely a matter of who can most efficiently utilize the available resources. In higher animals competition may involve direct interaction between individuals, as, e.g., through the establisment of breeding territories. An example of this was provided by Strandgaard (1972) who studied a population of roe deer and showed that the presence of very dominating males actually decreases the population density below the level that the productivity of the environment allowed.

To summarize the theoretical results, selection in a population at its equilibrium size, the carrying capacity, may favor both alleles which increase competitive ability and alleles which increase the reproduction potential. Thus, in its strict interpretation, both $r$- and $K$-selection may occur, but selection may even result in a decrease in the population size. On the other hand, $r$-selection may, whether it occurs in a population at equilibrium or during the initial increase of a population, result in an increase or a decrease in the number of offspring per individual. Thus $r$- and $K$-strategies may not be immediately reflected in simple life cycle characteristics of higher organisms, such as in the number of offspring. A better reflection of the two different strategies is contained in the total energy $E$ used for reproduction as compared to the energy spent for nonreproductive purposes, i.e., competition. This is the actual basis of the concept to many authors (e.g., Gadgil and Solbrig, 1972; Pianka, 1972).

The present findings, i.e., that selection in a competitive environment will not necessarily lead to a maximized equilibrium population size and that density independent selection may lead to an optimum litter size show that the widely used terms $r$- and $K$-selection are in some respects not fortunate.

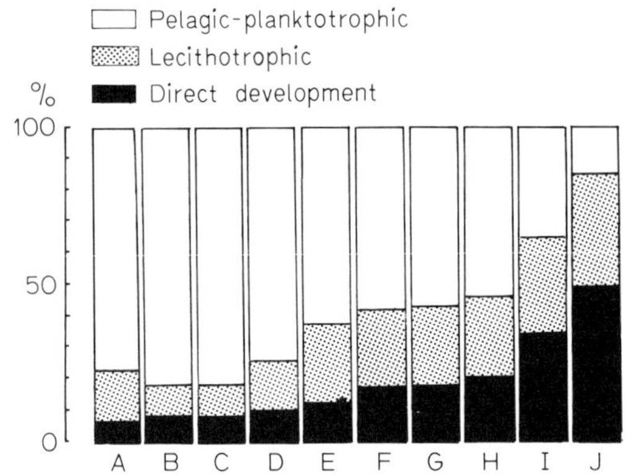


Fig. 14. The percentage of bivalves with different kinds of development along the shores of Europe from south to north. *A*: Gibraltar-Arcachon, *B*: Brest-St. Malo, *C*: the Channel, *D*: Southern North Sea, *E*: Northern North Sea and Danish waters, *F*: Stavanger, *G*: Lofoten, *H*: Western Finmark, *I*: Eastern Finmark, *J*: Spitzbergen. (After Ockelmann, 1965)

In this context a fruitful and still rather neglected area for theoretical population biology is the explanation of a large and fascinating body of observational data on the life histories of different animals and geographical distributions of reproductive patterns of related species. Cole (1954) in a pioneering paper studied the consequences of life length, spacing between, and sizes of litters and development time on $r_m$ using the theory developed in the preceding section. The results (which will not be reviewed here) apply to many observational data although all the models imply *r*-selection. There have so far been few attempts to follow this approach (e.g., Lewontin, 1965; Cohen, 1967; Mertz, 1971). Such studies should be extended to cover situations with competitive interactions and special biological adaptations such as parasitism and complex life cycles. However, the mathematical structure of models of density limited populations with continuous reproduction and in which demographic components and structure are included will be very complex and a general analysis such as the one carried out previously in this chapter under very simplified assumptions, may not be possible.

As an example of an interesting and largely unanswered problem we will mention the geographical distribution of reproductive adaptations in marine invertebrates. Some marine invertebrates produce *planktotrophic* larvae. These are small and feed in the plankton, and they are produced in huge numbers by each adult individual. Other forms have *lecithotrophic* larvae which are larger and provided with some nutrients. Finally some forms have direct development, in some cases in conjunction with brood protection. It is understandable that many littoral forms of isolated islands and species living in isolated estuaries do not have long-lived planktonic larvae. No satisfactory explanation, however, has so far been given the latitudinal pattern of reproduction (Thorson, 1950; Ockelmann, 1965); in the tropics nearly all benthic invertebrates have planktotrophic larvae, at higher latitudes an increasing number of species have lecithotrophic larvae; in the high arctic practically all species have direct development often in conjunction with brood protection (Fig. 14).

# 2. Interactions Between Species

Other plants and animals constitute an important part of the environment for all living organisms. Any species is a part of a web constituted by species which exploit each other or compete for vital resources. The persistence of a species in a habitat must consequently be understood from its position in the biotic community to which it belongs. All heterotrophic organisms provide for the energy and nutrients which are needed for maintenance and growth by assimilating and oxidizing organic molecules. This is frequently done by eating another organism; such relationships are called *prey-predator* relationships if the "victim" is smaller or of about the same size as the "profiteer" and *host-parasite* relationships if the victim is larger than the profiteer and the former is often not killed by the latter. This fundamental relationship of energy transfer from primary producers of organic matter (plants) through herbivores and detritivores to different levels of carnivores constitutes one dimension of a community often described in terms of *food chains*. The limited amount of energy available at each level of the community web gives rise to *competition* between species with overlapping requirements and this competition constitutes an alternative direction of relationships in the community often described in the terms of *trophic levels*.

A theoretical study of complex communities is naturally initiated by a description and classification of relationships between pairs of interacting species. Following the terminology of Odum (1971) the main interaction types are the mentioned prey-predator (− +) and competitive (− −) relationships to which mutualism (+ +), commensalism (○ +), ammensalism (○ −), and the trivial indifference (○○) can be added. These, however, are only the bricks of a qualitative description of a community; in order to arrive at predictive models, quantification of the intensity of the interactions is needed.

A community of $m$ interacting species may be described by a set of equations like

$$dx_i/dt = x_i F_i(x_1, x_2, \ldots, x_i, \ldots, x_m) \tag{2.1}$$

where $x_i$ quantifies the population size of the $i$'th species and $F_i$ describes the relative change in population size of the $i$'th species as a function of the population sizes of the species in the community. The functional form of the $F_i$'s contains the qualitative and quantitative description of the community interactions. In the following we will explore some simple models of the form [Eq. (2.1)] that are more or less open to analytical investigations.

When aiming at an analysis of a general community model as symbolized by Eq. (2.1), it is important to have a profound knowledge of the properties of the fundamental interactions in the system, namely the interactions between pairs of species. These interactions constitute the atoms in the web of interactions of the whole community. Also, two species systems are open to a more complete analysis than are the more complex multispecies systems.

## 2.1 One-Prey–One-Predator Models

Lotka (1925) and Volterra (1926) were the first to formulate and analyze a simple two-linked food chain. Independently they proposed the following simple differential equations to describe a predator-prey system;

$$dx/dt = x(r - \gamma_1 y), \tag{2.2a}$$

$$dy/dt = y(-d + \gamma_2 x), \tag{2.2b}$$

where $x$ and $y$ are the population sizes of the prey and the predator respectively, $r$ is the unlimited growth rate of the prey species in the absence of the predator, $d$ is the death rate of the predator, and $\gamma_1$ and $\gamma_2$ are constants describing the loss of the prey and the gain of the predator respectively. The model is based on a series of simplifying assumptions, the most important of which are the following.

1. The prey is only limited by the predator; in the absence of the predator the prey population would grow unlimited. The model is thus describing a situation where the limiting factor for the prey species is predation and not intraspecific interaction.

2. The number of prey eaten per time unit is proportional to random encounters between individuals of the prey and the predator populations. This is an unrealistic functional response of the predator to its prey (Holling, 1965; Maynard Smith, 1974) since it infers that the predator can eat unlimited amounts and does not reach any state of satiation.

3. The predators only compete for prey and no other limiting resources (e.g., space) are allowed for.

4. All responses of the populations are instantaneous; the model does not allow for time-lags, e.g., the time it takes the predator to handle and ingest the prey.

5. Finally, like most simple ecological models with the scope of giving general insight, it ignores age structure and other time correlations of the populations.

The Eqs. (2.2a and b) yield one equilibrium point with $\hat{x} = d/\gamma_2$ and $\hat{y} = r/\gamma_1$ where both derivatives vanish and the population sizes remain constant. Dividing Eq. (2.2a) by Eq. (2.2b) and integrating with respect to $x$ and $y$ yields the function

$$Q(x, y) = r \log y + d \log x - \gamma_1 y - \gamma_2 x.$$

This function is constant in time ($dQ/dt = 0$), and its value is dependent on the initial population sizes $x_0$ and $y_0$. The equation $Q$ = constant determines a closed curve which describes an oscillating system. The system [Eqs. 2.2a and b)] thus

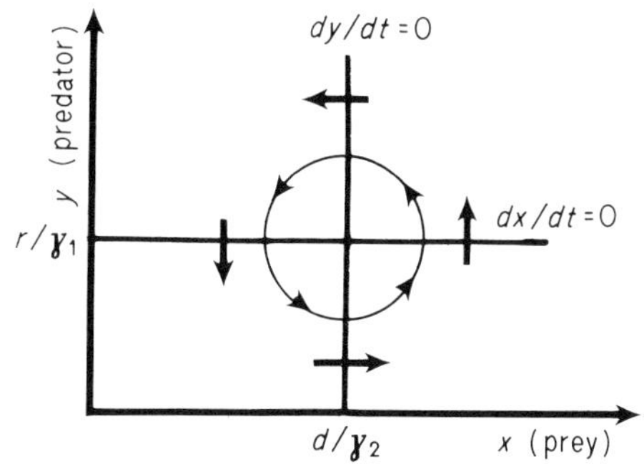


Fig. 15. A graphical analysis of model [Eq. (2.2)]

shows undamped oscillations with neutral stability; any pertubation of the populations will only change the population trajectory to alternative amplitude values. It can formally be shown (Pielou, 1969) that the average population sizes of $x$ and $y$ in all cases are the equilibrium values and that the oscillations of the two populations are 1/4 cycle out of phase and the period is independent of amplitude and equals $2\pi/\sqrt{rd}$. A graphical analysis of the equations in the $(x, y)$ plane is shown in Figure 15.

The undamped, neutrally stable oscillations of the prey-predator system [Eqs. (2.2a and b)] are a consequence of the special and very simple structure of the model. The only density regulating mechanisms of the model are interspecific and this results in the time-lags of the reactions. These pathological characteristics disappear as soon as intraspecific competition is allowed for in the model. Volterra (1926) proposed the following simple version of the model;

$$\begin{aligned} dx/dt &= x(r - \alpha x - \gamma_1 y)\,, \\ dy/dt &= y(-d + \gamma_2 x)\,, \end{aligned} \tag{2.3}$$

in which density dependent regulation of the prey species is considered. This model is analyzed graphically in Figure 16. Another extension of the model is, of course, to allow the predators to compete too, thus giving the general model

$$\begin{aligned} dx/dt &= x(r - \alpha_1 x - \gamma_1 y)\,, \\ dy/dt &= y(-d + \gamma_2 x - \alpha_2 y)\,. \end{aligned} \tag{2.4}$$

Figure 17 shows and analysis of this model and one described by Leslie (1948) in which the functional form of the prey-predator interaction is changed relative to Eq. (2.2). In this model the predator is limited by the number of prey per predator rather than by the number of prey;

$$dy/dt = y[s - y/(\gamma_2 x)]\,. \tag{2.5}$$

This model corresponds to the logistic model Eq. (1.3) with the carrying capacity as a simple function of the prey abundance.

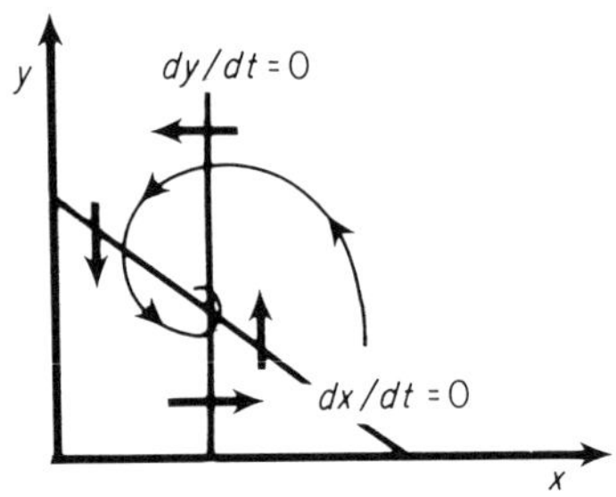


Fig. 16

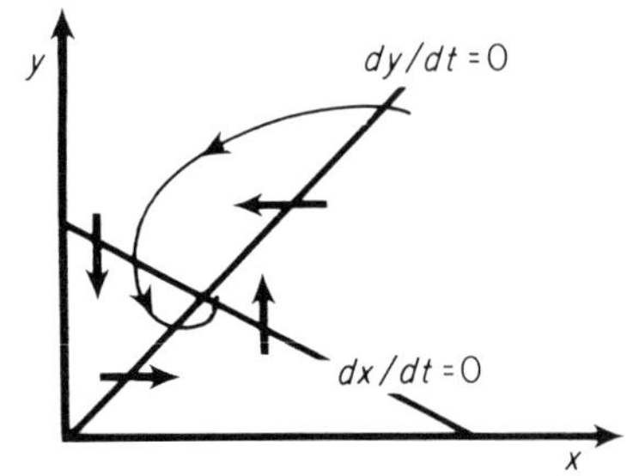


Fig. 17

Fig. 16. A graphical analysis of model [Eq. (2.3)]

Fig. 17. A graphical analysis of models [Eq. (2.4) and Eq. (2.5)]

Kolmogoroff (1936) (see also Rescigno and Richardson, 1967) devised a general prey-predator model formulated by

$$dx/dt = xF(x, y),$$
$$dy/dt = yG(x, y), \tag{2.6}$$

where the functions $F$ and $G$ have the following six properties:

1. When both species are rare the prey will increase, i.e., $F(0, 0) > 0$.
2. The per capita increase of the prey population is a decreasing function of the size of the predator population, i.e., $\partial F/\partial y < 0$.
3. The multiplication of the predators decreases with their number, i.e., $\partial G/\partial y < 0$.
4. For a constant ratio of prey to predators, the multiplication is slowed by an increase in the number of predators, i.e., $x(\partial F/\partial x) + y(\partial F/\partial y) < 0$, and the multiplication of the predators is enhanced by an increase in the prey, i.e., $x(\partial G/\partial x) + y(\partial G/\partial y) > 0$. These properties reflect that encounters between prey and predators become more frequent with higher population densities.
5. There exists a critical size $b > 0$ of the prey population below which the predator cannot increase, i.e., $G(b, 0) = 0$.
6. High predator densities or high prey densities prohibit prey growth, i.e., there are constants $A$ and $a > 0$ such that $F(A, 0) = F(0, a) = 0$.

It may now be shown (Kolmogoroff, 1936) that when 1. through 6. are fulfilled, then the model describes a stable limit cycle or a stable equilibrium if and only if $b < A$ (Fig. 18). The result does not provide conditions for whether a stable equilibrium or a stable limit cycle will be the solution. A local analysis of the equilibrium $E$ (Fig. 18) does not in general make that distinction because it is possible for $E$ to be stable and at the same time be surrounded by a stable limit cycle. However, when $A$ (the carrying capacity of the prey) is much larger than the size of the prey population at the equilibrium $E$, i.e., when the model has properties approaching those of the Lotka-Volterra model Eq. (2.2), then we may expect a stable limit cycle. Conversely, if $A$ and the equilibrium prey population are not too different, the equilibrium $E$ becomes globally stable.

The general model [Eq. (2.6)] may be used to analyze a variety of models (May, 1973a) and in many situations it may support a global analysis. The simple

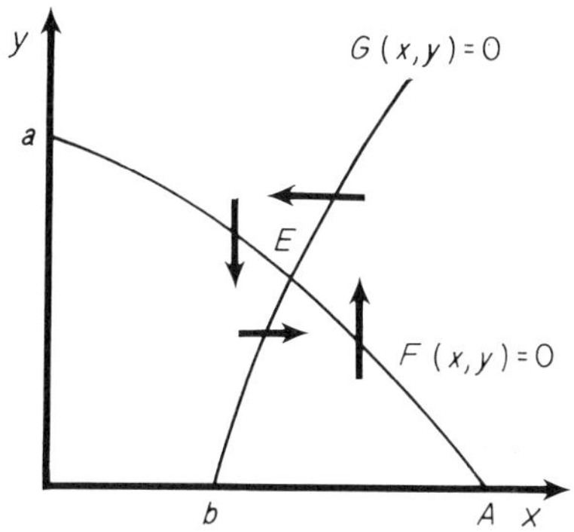


Fig. 18 Fig. 19

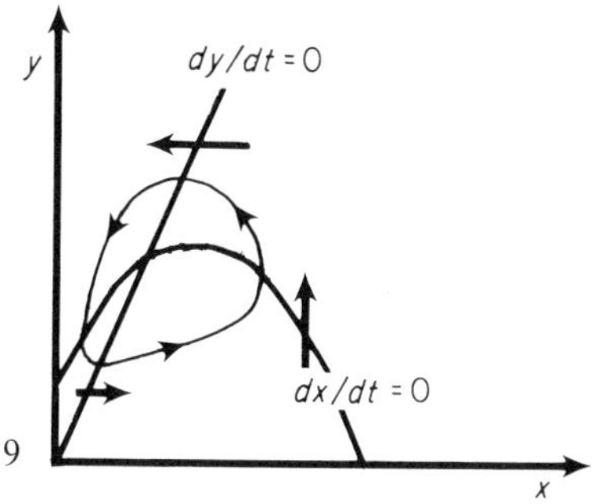


Fig. 18. A graphical representation of the general prey-predator model (2.6)
Fig. 19. A prey-predator model showing a stable limit cycle. (After May, 1973a)

model [Eq. (2.4)] is such an example. It fulfills all the requirements for stability except condition 4. on the function $G$. This condition is only fulfilled for $\gamma_2 x > \alpha_2 y$, i.e., there are difficulties for excessive predator densities. However, in that region of the $(x, y)$ plane the requirement for stability is that the predator decreases fast enough to prevent the prey from going extinct, but condition 4. on $G$ is a requirement that the predator increases fast enough at low predator densities.

The condition should, therefore, not be needed in that particular region and in fact it may be shown that if the population starts here it will go out of the region while being bounded away from the $x=0$ edge (see below). Therefore, the prey population will never go extinct due to excessive predator densities but will always return to an area where the requirements for Kolmogoroff's result are fulfilled and neither of the species will go extinct. In general Kolmogoroff's results apply if the isoclines have an appearance as in Figure 18, specifically if they only intersect in one point.

An example of a model fulfilling the requirements 1. through 6. is the model [Eq. (2.5)]; this supports the graphical analysis of the model (Fig. 17).

An example of a predator-prey model which exhibits a stable limit cycle was given by May (1973a) with a model which is a hybrid between the models of Leslie (1948) given by Eq. (2.5) and the model of Holling (1965):

$$\begin{aligned} F(x, y) &= r[1 - x - y/(x+\delta)]\,, \\ G(x, y) &= s[1 - y/(\gamma x)]\,. \end{aligned} \tag{2.7}$$

Kolmogoroff's results apply to this model though not all the requirements 1. through 6. are completely fulfilled. Therefore, the predator-prey system is always globally stable; when the equilibrium point is unstable there must exist a stable limit cycle (Fig. 19).

A general type of prey-predator model is the following which is an extension of a model considered by Rosenzweig and MacArthur (1963):

$$\begin{aligned} F(x, y) &= r(x) - \gamma_1 y \varphi(x)\,, \\ G(x, y) &= -d(y) + \gamma_2 x \varphi(x)\,, \end{aligned} \tag{2.8}$$

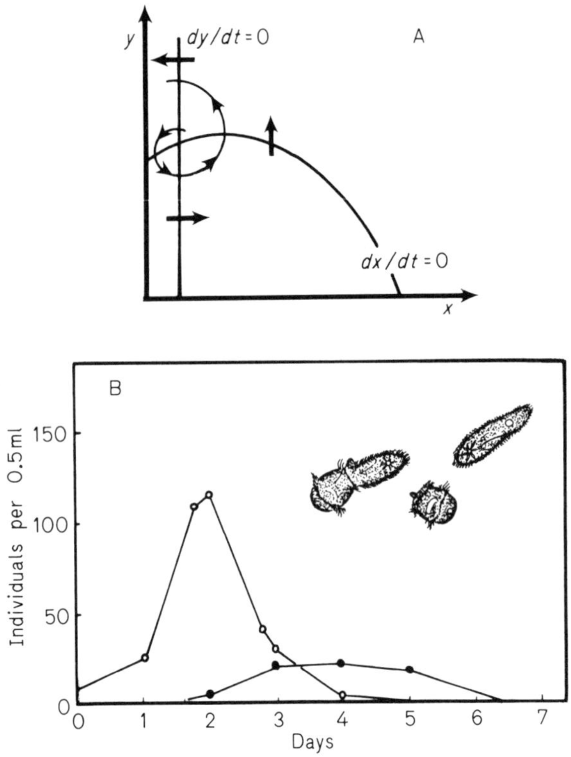


Fig. 20A and B. Examples of unstable prey-predator systems. (A) a theoretical model. (B) Population sizes of *Paramecium caudatum* (○) and its predator *Didinium nasutum* (●) in an experimental system. (After Gause, 1934)

where $r(x)$ is the density dependent growth rate of the prey in the absence of the predator, $d(y)$ is the death rate of the predator assumed to be density dependent, and $\varphi(x)$ is the rate of predation per prey-predator encounter which is allowed to be dependent on prey abundance.

The functions describing the model must be reasonable. We may assume that the growth rate of the prey, $r(x)$, is a decreasing function of $x$, that the death rate of the predator, $d(y)$, is an increasing function of $y$, and that the predation rate per predator, $x\varphi(x)$, is an increasing function of the prey density, $x$. By these assumptions requirements 1. through 3. and the requirements with respect to $F$ in 4 of Kolmogoroff's theorem are fulfilled. By additional, reasonable constraints, requirements 5 and 6 are also fulfilled. As for the Lotka-Volterra model, the only trouble is requirement 4 with respect to function $G$. This condition now reads

$$\gamma_2 x \partial[x\varphi(x)]/\partial x - y\partial d/\partial y > 0 \tag{2.9}$$

which is not fulfilled for large values of $y$. However, for sufficiently small $y$ it is always fulfilled so that the predator cannot go extinct while the prey population persists, i.e., the trajectory of the populations cannot hit the $x$-axis. The question

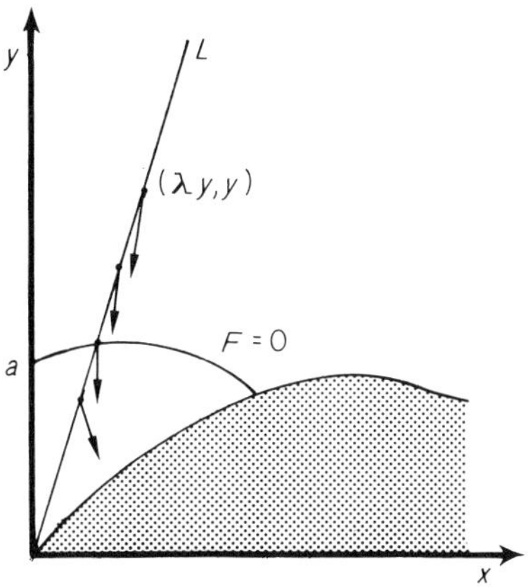


Fig. 21

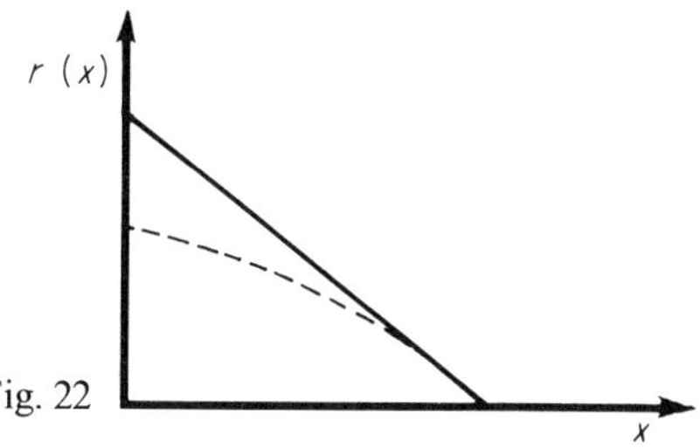


Fig. 22

Fig. 21. The population trajectories in points on the line $L$ in a predator-prey system of the type described by Eq. (2.8). In the shaded area below the isocline $x(\partial G/\partial x)+y(\partial G/\partial y)=0$, Kolmogoroff's theorem is always valid. For further explanation, see text

Fig. 22. The growth rate as function of density as assumed by the model [Eq. (2.4)] and (*interrupted line*) allowing for a relatively lower growth rate at low densities

is now whether the trajectory can hit the $y$-axis. We will show that under wide conditions there exists a line $L$, given by $x=\lambda y$, such that all trajectories that pass this line or any line steeper than $L$ will pass in the direction away from the $y$-axis; see also Figure 21.

The slope of a trajectory is $[yG(x, y)]/[xF(x, y)]$. For points on $L$, where $x=\lambda y$, the slope will be given by

$$(1/\lambda)\, G(\lambda y, y)/F(\lambda y, y)\,. \tag{2.10}$$

The condition that the trajectory is steeper than the line is that Eq. (2.10) is greater than $(1/\lambda)$ or that $G/F$ is greater than one. Remembering that we are only interested in the condition for $F<0$ (for $F>0$ the trajectory is evidently away from the line $L$ and the $y$-axis) then we get the condition

$$G(\lambda y, y)<F(\lambda y, y)$$

which for Eq. (2.8) yields

$$(\gamma_2\lambda+\gamma_1)\, y\varphi(\lambda y)<r(\lambda y)+d(y)\,.$$

This condition is fulfilled for all $\lambda$ sufficiently small if

$$y\gamma_1\varphi(0)<r(0)+d(y) \tag{2.11}$$

for all $y>a$. Thus if the death rate of the predator increases sufficiently fast with its density, then it cannot eat up the prey. Comparing Eqs. (2.11) and (2.8) reveals that the condition is that with a small prey population, the predation pressure on the per capita growth rate of the prey should increase less with the predator density than should the per capita death rate of the predator.

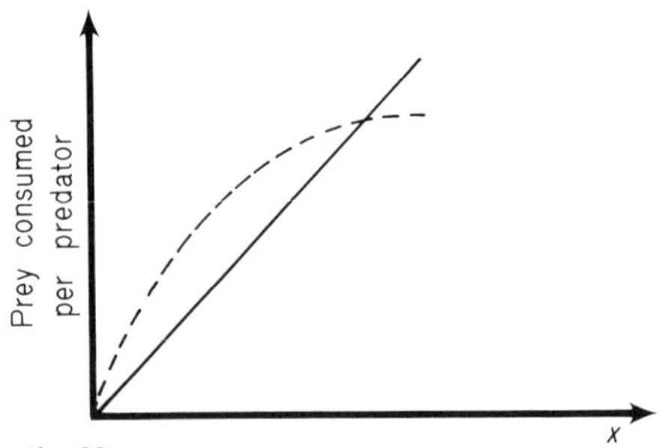


Fig. 23

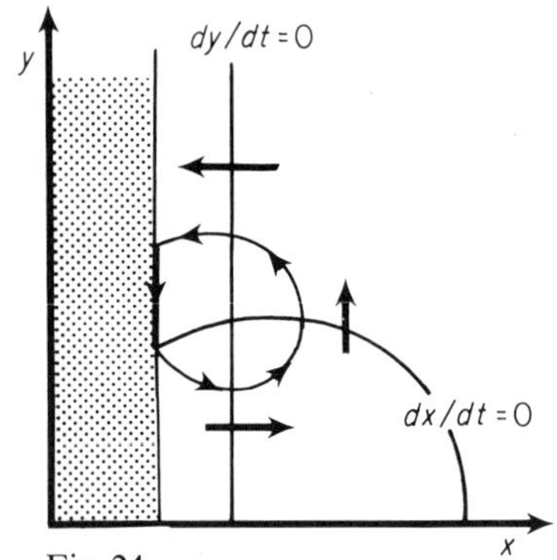


Fig. 24

Fig. 23. The functional response of a predator to prey density as assumed by the model [Eq. (2.4)] and (*interrupted line*) allowing for a saturation effect

Fig. 24. A predator-prey system as shown on Figure 20A but with the prey having a refuge which stabilizes the system

This condition for global stability of the model [Eq. (2.9)] is only a sufficient condition; as mentioned earlier, the simple model [Eq. (2.4)] is always globally stable even though Eq. (2.11) is not fulfilled. This may be shown by using the same technique as above but substituting the line $L$ with a parabola, $x=\lambda y^p$, where $p>\gamma_1/\alpha_2$.

The model [Eq. (2.8)] with the above mentioned conditions covers a wide range of effects neglected by the simple model [Eq. (2.4)]. The growth potential of the prey may be low at low densities due to difficulties in finding mates or for other reasons (Fig. 22). Similar density dependent effects on the death rate of the predator are allowed. Furthermore the model allows for a general saturation effect in the predator's response to the prey (Fig. 23 and Holling, 1965) and for the prey having a refuge from the predator (Fig. 24) which always stabilizes the system.

Apart from the interest in global stability of a predator-prey system, it is also of interest to know whether the system will stabilize at the equilibrium $E$ (Fig. 18) or whether it will exhibit a stable limit cycle. As already mentioned, conditions for global stability of $E$ have not been found, but we may ask for conditions that $E$ is locally stable, i.e., if the populations are started within a specified circle around $E$ they will then converge to $E$. This is more or less the condition that if the system is started with population sizes corresponding to $E$ it will stay there, i.e., it becomes a necessary condition for $E$ being a globally stable equilibrium. A local stability analysis is in principle carried out as done in Chapter 1; a general framework for the analysis of local stability will be discussed later in this chapter. Here we will just mention that equilibria where the $F=0$ and the $G=0$ isoclines intersect as shown in Figures 17 and 18 are always locally stable (qualitatively stable in the sense of May, 1973a), and equilibria where the isoclines intersect like in Figure 19 may or may not be stable depending on the parameters. The condition for qualitative stability is thus that the slope of the $F=0$ isocline at $E$ is negative and that of the $G=0$ isocline is positive, and

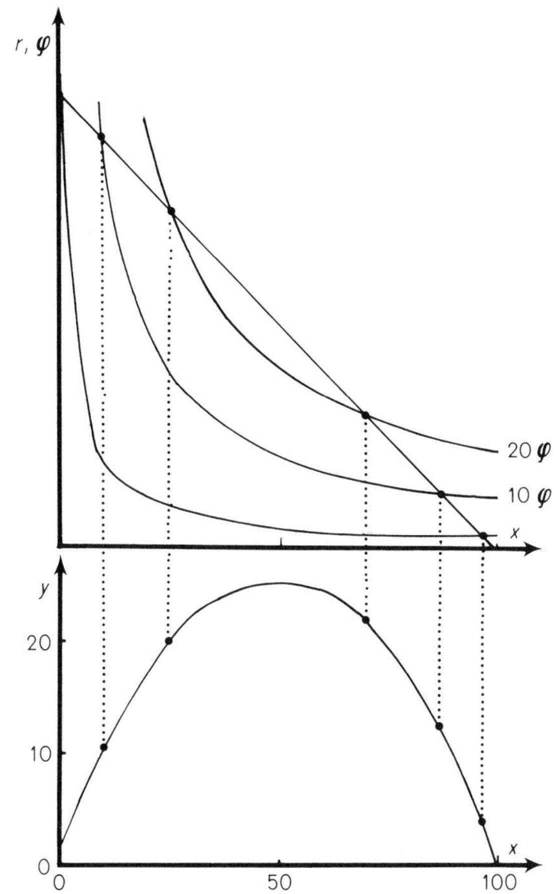


Fig. 25. The graphical construction of an $F=0$ isocline with a "hump" due to a saturation of the predator with prey. Each point of the curve is found as the intersection of the $r(x)$ function and the functions $y\varphi(x)$. When there are two intersections, the $F=0$ curve will have a hump. The growth rate of the prey is of logistic form, i.e., $r(x)=r(K-x)/K$, and the response of the predator to prey density is of the form shown in Figure 23, i.e., $x\varphi(x)=bx/(x+a)$

when both are positive the stability depends on the parameters. This latter possibility is probable when the $F=0$ has a "hump" or maximum; in the case of the model [Eq. (2.8)] this is possible when the predator can be saturated with the prey as demonstrated graphically in Figure 25.

When the equilibrium $E=(\hat{x}, \hat{y})$ is unstable and the populations stabilize in a limit cycle, then the ratio between the minimum and maximum prey population size is approximately $\exp[-(cA/\hat{x})^2]$, where $c$ is close to unity (May, 1973a). Thus if $\hat{x}$ is much smaller than $A$ (the carrying capacity of the prey) the population size of the prey may become very low during some part of the cycle, in fact $<1$. This is relevant to experimental laboratory systems, where the extinction of the

Fig. 26. A simple, linear food chain

prey, even when its population minimum is larger than unity, is likely due to stochastic events.

However, the general picture which emerges from studying simple predator-prey systems is that they act as stabilizers of the community web if both the prey and predator populations are density regulated extrinsic to the prey-predator relation. This property may be extended to simple linear food chains (Fig. 26) as discussed by May (1973a). However, such food chains must be restricted in length and cannot themselves contribute much to the diversity of biotic communities. At each link of the food chain only a fraction of the assimilated energy can be used for growth and reproduction, the remaining is used for maintenance. Further, not all food which is ingested is assimilated. We may take 10% as a crude estimate of the amount of energy ingested by one link in the food chain which can be passed on to the next link (Slobodkin, 1961). Thus in a food chain with say 5 links, the "top carnivore" will only receive about $10^{-5}$ times the energy available to the first link, e.g., green plants. In nature, food chains are in fact rarely longer than 5 links.

Several experimental predator-prey systems are described in the literature. This approach, i.e., to study simple microcosms with for example populations of a prey species and its predator and compare their behavior with simple mathematical models may be of value for clarifying properties not built into the models (e.g., special behavioral properties of the species in question). This type of experiment, however, cannot "prove" or "disprove" simple population models. If we carefully select organisms and experimental conditions which conform with the model, then we can at most prove that the theoretician is capable of integrating his equations; using a real biological system only for simulation of a model does not yield more insight about the real system or the model than we had before the experiment, and we could as well have used an electronic computer.

Many of the experimental systems described in the literature show divergent oscillations, i.e., they are unstable. Most known are the experiments of Gause (1934, 1935) who used the ciliate *Paramecium caudatum* and its ciliate predator *Didinium nasutum*. Gause's systems invariably went extinct after 1–2 oscillations (Fig. 20) unless low, constant migration rates of both species were imposed on the systems. The main reasons for the divergent oscillations are time-lags in the response of the predator to decrease in the prey population (*Didinium* may continue to divide after starvation, producing smaller individuals, and it thus violates requirement 5 of Kolmogoroff) in conjunction with small total population sizes so that the trajectories came fatally close to the $y$- or the $x$-axis (Williamson, 1972; Maynard Smith, 1974). Luckinbill (quoted from Maynard Smith, 1974) found that adding methyl cellulose to the medium (thus increasing the viscosity of the medium which decreased the numbers of prey-predator encounters) or decreasing the food level of the prey rendered Gause's systems stable. As discussed above, smaller carrying capacities of the prey species tend to damp the oscillations.

Utida (1957) was able to sustain oscillating populations of a bean weevil and a parasitoid wasp for more than 100 generations. Realistic models of systems which involve higher animals with complex life cycles and behavior patterns must include age structure and time-lags and are mainly based on difference equations. Such models are outside the scope of this book; the reader is referred to Nicholson and Bailey (1935), Maynard Smith and Slatkin (1973), Maynard Smith (1974), and Usher and Williamson (1974). Prey-predator systems in a heterogeneous environment and with migration between patches tend to be more stable than the same systems in a homogeneous environment (cf. Chap. 4). Huffaker's (1958) experiment exemplifies this. The systems consisted of oranges with a herbivorous mite and its predator, another mite. Many of these systems went extinct after some oscillations but when the spatial arrangement of the system, i.e., trays with oranges, became sufficiently complex including migration barriers, etc., the systems showed stable limit cycles. This system is analyzed by Maynard Smith (1974).

The well-known oscillating hare-lynx system in Canada may, together with a few others, be taken as an example of stable limit cycles of a prey-predator system in nature. As pointed out by May (1973a) this system cannot, however, be taken as an illustration of the neutral oscillations of the Lotka-Volterra model [Eq. (2.2)] as inferred in some text books. May (1973a) also quotes an example where a natural prey-predator system seems to show stable limit cycles where conditions are favorable for the prey species but shows a stable equilibrium where conditions are less favorable; i.e., the decrease of the $K$-value of the prey damps the oscillations in accordance with the theoretical predictions.

## 2.2 Two Species Competing for Resources

Lotka (1925) and Volterra (1926) also considered the case when two species compete for the same resource and proposed the following equations:

$$\begin{aligned} dx_1/dt &= r_1 x_1 (K_1 - x_1 - \beta_{12} x_2)/K_1\,, \\ dx_2/dt &= r_2 x_2 (K_2 - x_2 - \beta_{21} x_1)/K_2\,, \end{aligned} \tag{2.12}$$

where $K_i$ are the carrying capacities of the two species when living alone in the habitat and $\beta_{ij}$ ($\geqq 0$) are the *competition coefficients* which measure the degree of inhibition of species $i$ due to the presence of species $j$. Clearly if $\beta_{12} = \beta_{21} = 0$ there is no interaction between the species and Eq. (2.12) degenerates into two

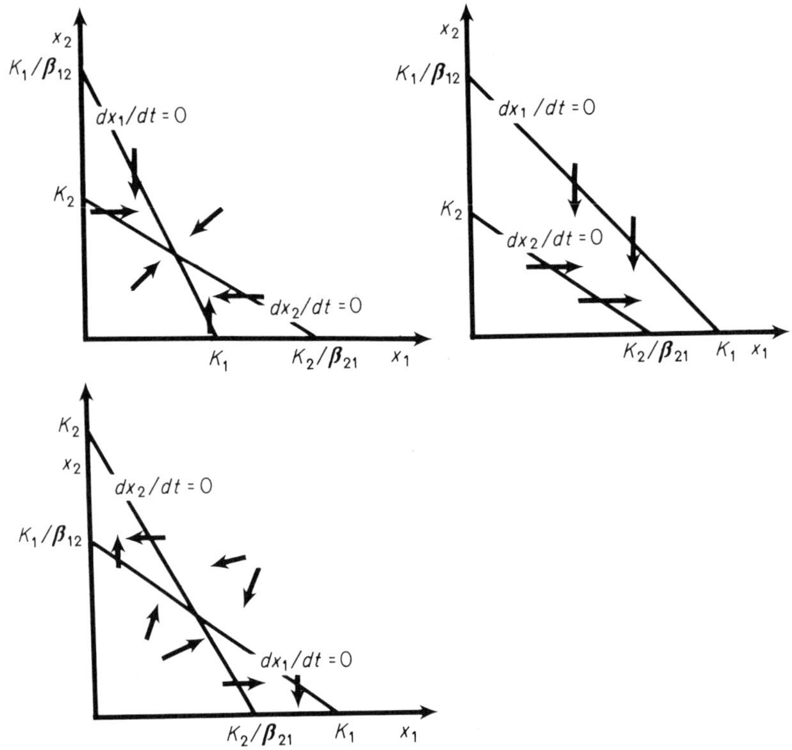


Fig. 27. A graphical analysis of a two species competition system obeying model [Eq. (2.12)]. *Upper left*: Situation with stable coexistence. *Upper right*: Situation where $x_1$ always wins. *Lower*: Situation where the final outcome of competition is determined by the initial population sizes

logistic equations. If $x_1$ and $x_2$ are the absolute numbers of the two species, then $\beta_{12}$ measures the inhibition of species 2 on species 1 in units of the inhibition of species 1 on its own growth. However, the species abundance may be measured in any unit so we can choose to measure the size of the population of species 1 in units of its inhibition on species 2 (when $\beta_{21} > 0$); this coordinate transformation changes [Eq. (2.12)] into the equivalent system;

$$\begin{aligned} dx_1/dt &= r_1 x_1[(K_1\beta_{21}) - x_1 - \beta_{12}\beta_{21}x_2]/(K_1\beta_{21}), \\ dx_2/dt &= r_2 x_2(K_2 - x_2 - x_1)/K_2 . \end{aligned} \tag{2.13}$$

Therefore, the competition between the two species is completely characterized by the quantity $\alpha = \beta_{12}\beta_{21}$. If the two species have complete overlap in the utilization of resources then the inhibition in growth is for both species determined by the quantity $x_1 + x_2$ and $\alpha = 1$. When the overlap in resource utilization is only partial then $\alpha < 1$, and when the two species inhibit each other in other ways than just by using the same resources, e.g., by preying on each other or by other significant antagonistic mechanisms, then $\alpha > 1$.

The qualitative consequences of Eq. (2.12) can be analyzed graphically by observing the signs of $dx_i/dt$ and thereby the direction of movement of the population in the $(x_1, x_2)$ plane. The changes in signs occur along the $dx_i/dt=0$ isoclines, i.e., the straight lines $K_i - x_i - \beta_{ij}x_j = 0$. As seen in Figure 27 there are four qualitatively different outcomes of Eq. (2.12) according to the configurations of the isoclines. The four possibilities are as follows: (1) Species 1 wins independent of initial population densities. (2) Species 2 wins independent of initial population sizes. (3) Either species 1 or species 2 wins according to initial population densities. (4) Both species remain in the system in stable coexistence. The exact requirement for stable coexistence is that $\beta_{ij}/K_i < 1/K_j$ which can be interpreted to mean that each species inhibits its own growth more than it inhibits the other species. In the special case where $K_1 = K_2 = K$ it can be seen that $\beta_{12}$ and $\beta_{21} < 1$ is a necessary and sufficient condition for coexistence. In this case, the total population size at equilibrium, $\hat{x}_1 + \hat{x}_2$, exceeds $K$ which corresponds to the interpretation that there is not a total overlap in resource utilization in this case ($\beta_{12}\beta_{21}$ is necessarily less than unity). This result would seem to indicate that two species may be as similar as conceivable save for complete overlap in the resource utilization and still coexist. The case where either of the species can win dependent on initial conditions, on the other hand, requires that both $\beta_{ij}$'s are greater than unity, i.e., that some antagonistic mechanism is involved.

The graphical analysis described above was extended by Rescigno and Richardson (1967) to cover models of a general structure like the Eqs. (2.6) describing a predator–prey system. The functions $F$ and $G$ are now specified according to the qualitative properties of competition. In the present context the model should be formulated as

$$dx_i/dt = x_i F_i(x_1, x_2) \quad \text{for} \quad i = 1, 2\,. \tag{2.14}$$

The requirements for a global analysis are the following assumptions.

1. Each species will increase if both are rare, i.e., $F_i(0,0) > 0$ (a simplifying assumption which is not very crucial; an Allee-effect could be allowed).

2. The per capita increase of each population is a decreasing function of its population size and the population size of the other species, i.e., $\partial F_i/\partial x_j < 0$ for $i$ and $j = 1, 2$.

3. For each species when alone, there exists a critical population size above which it cannot increase, i.e., $F_1(A_1, 0) = F_2(0, A_2) = 0$ where the $A$'s (i.e., the carrying capacities) are positive.

4. For each species when alone, there exists a critical population size above which the absent species cannot increase, i.e., $F_1(0, a_1) = F_2(a_2, 0) = 0$ where the $a$'s are positive.

An example is shown in Figure 28. Here the $F_i(x_1, x_2) = 0$ isoclines intersect in two points and an inspection of the directions of movements as specified by the signs of $F_1$ and $F_2$ show that $S$ and $A_1$ are stable equilibria. The domains of attraction of these two points cannot be found explicitly without a more specific definition of the functions $F_i$. However, the dotted lines through the unstable equilibrium $U$ partition the $(x_1, x_2)$-plane into four regions where the one containing $S$ is contained in the domain of attraction of $S$ and similarly for

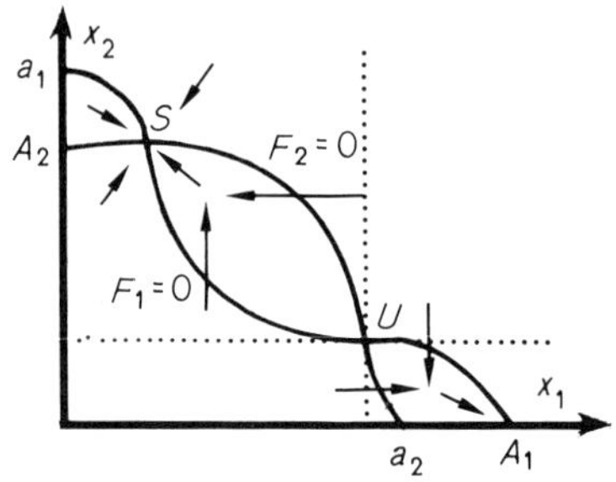


Fig. 28. A two species competition system: An example of a general case with two stable equilibria, viz., $S$ and $A_1$

the region containing $A_1$ on the edge. Observe that the equilibria $S$, $U$, and $A_1$, when viewed locally correspond to the situations of the Lotka-Volterra model [Eq.(2.12)] shown in Figure 27. Thus in this way the simple linear model [Eq.(2.12)] may be viewed as an approximation to the general model [Eq.(2.14)] when considering populations close to an equilibrium.

It is clear from Figure 28 that a stable coexistence is guaranteed when $A_i < a_j$ for $i \neq j$, a condition formally equivalent to the condition of the Lotka-Volterra model. The other property of the Lotka-Volterra model, that the total population size of the combined system is larger than the carrying capacity of each species does not necessarily generalize to any model (Maynard Smith, 1974).

Many experimental laboratory experiments illustrate the qualitative predictions of models such as [Eqs.(2.12) or (2.14)]. As in the case of similar predator-prey systems, the analysis of simple experimental systems may yield insight in the particular species used or suggest new general principles, but they cannot prove or disprove simple mathematical models.

The outcome of competition experiments always shows that when two species are forced to share a common, limiting resource, one of the species will be excluded unless the two species can use some heterogeneity of the experimental universe and thereby share the resource through habitat selection. Gause (1934, 1935) carried out competition experiments with different species of *Paramecium* competing for bacteria or for yeast cells; these food items were supplied to the systems at a constant rate. Experiments with *Paramecium caudatum* and *Paramecium aurelia* always ended with the competitive exclusion of one of the species according to the experimental conditions. In these experiments 10% of the populations were removed every day for enumeration and then discarded. Consequently the populations were subject to a density independent mortality rate so that the growth rates of the species played a role for the outcome of the experiments. In experiments with *P. caudatum* and *Paramecium bursaria* competing for yeast cells in tubes, a stable coexistence was found. This could formally be described in terms of the Lotka-Volterra equations in the case of stable coexistence ($\beta_{ij}/K_i < 1/K_j$). However, the underlying mechanism for the coexistence is that *P. bursaria* is a superior competitor in the sedimented yeast cells in the bottom of the tubes and is able to exclude *P. caudatum* while *P. caudatum* on the other hand is able to exclude *P. bursaria* in the yeast suspension above the bottom.

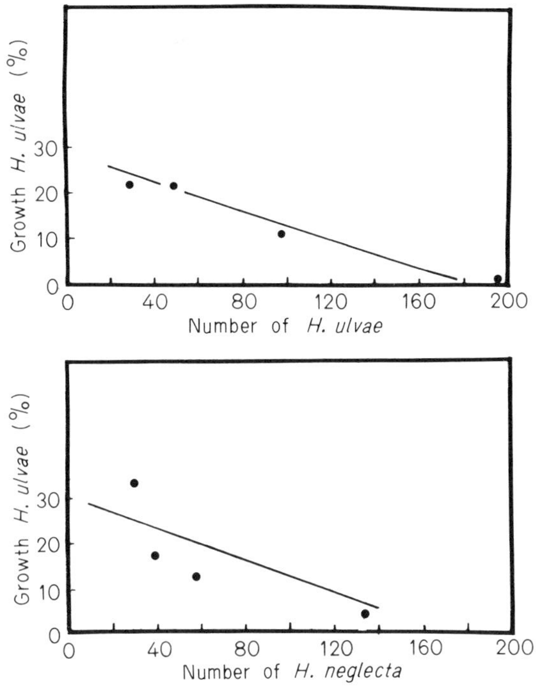


Fig. 29. The individual growth during two months in about 20 $cm^3$ sediment of the mud snail *Hydrobia ulvae* as function of its own density in monocultures and as function of the density of snails in mixed cultures with *H. neglecta* containing 20 *H. ulvae*. (After Fenchel and Kofoed, 1976)

The two species therefore show habitat selection in a heterogeneous environment and each of the species is excluded from a part of their "fundamental niches" (Chap. 3). In an experimental design which secured a totally homogeneous environment one of the two species would always become excluded.

Vandemeer (1969) studied the interspecific competition of four species of ciliates for which the *K*'s and the $\beta$'s had been estimated by growing each of the species separately and in pairs. The outcome of the four-species system could then largely be predicted from a four-species version of Eq.(2.12).

Crombie (1946) studied the competition between different species of flour insects and also showed that spatial heterogeneity (e.g., small glass tubes into which only one of the competing species could enter) may lead to coexistence of species which cannot coexist in a homogeneous environment. In competition experiments with flour insects such as *Tribolium* spp. (e.g., Park, 1962) the interaction is in part antagonistic due to mutual predation on larvae and pupae ($\beta_{12}\beta_{21} > 1$) leading to the case where the outcome of the competition depends on the initial population sizes.

It is clear that when $K_1 \simeq K_2$ and $\beta_{12} \simeq \beta_{21} \simeq 1$, then the process of competitive exclusion or the arrival at an equilibrium situation is very slow; this is often the case with competition experiments with different species of *Drosophila* (e.g., Moore, 1952; Ayala, 1969; Barker, 1973).

The nature and quantification of competition will be discussed in detail in Chapter 3 with special reference to field conditions. In experimental populations the coefficients of competition may in principle be measured as the decrease of fitness of species 1 as a function of increasing numbers of species 2 measured by the intraspecific competition of individuals belonging to species 1. Such an experiment with two species of mud snails (*Hydrobia* spp.) is shown in Figure 29. These snails ingest diatoms and sediment particles and utilize the attached microflora of the latter. In containers with regularly exchanged marine sediment they show exploitative competition. On the graph the individual growth of *H. ulvae* is shown as a function of its own density (in about 30 cm$^2$ dishes with about 20 cm$^3$ sediment) and as the function of the density of the related species *H. neglecta* (in similar dishes and with 20 individuals of *H. ulvae*). It can be seen that $\beta_{ulvae,\,neglecta}$ is not very far from unity.

## 2.3 Mutualistic Pairs of Species

The species interactions not specifically discussed in the previous sections are commensalism (○+), ammensalism (○−), and mutualism (++). The two former may be viewed as special cases of the prey-predator and the competition models respectively. However, they degenerate to independent logistically growing species with time, depending on the carrying capacity of one of the species so that the results of Chapter 1 apply. Stable coexistence is thus guaranteed if both species always have positive "carrying capacities."

Mutualism, however, is a special quality of interaction which has rarely been considered from the point of view of theoretical ecology. This is regrettable since a vast number of such relationships, many of which are still incompletely understood, are known from nature. One need only to think about insects as pollinators, symbiotic microalgae in, e.g., corals and symbiotic nitrogen fixation to recognize the significance of mutualistic relationships. Besides the mentioned examples which are fundamental to the function of the biosphere, there are great numbers of more exotic examples, e.g., cleaning symbiosis of coral fish, hermit crabs with symbiotic actinians and fish and cephalopods with symbiotic, luminescent bacteria.

A simple linear model of mutualism would be

$$dy_i/dt = r_i y_i (K_i - y_i + \beta_{ij} y_j)/K_i \quad \text{for} \quad i = 1, 2\,. \tag{2.15}$$

However, some restraints must necessarily be imposed on the model. The benefit that one species gets from its symbiont can never be larger than the energy contained in the symbiont or in the resources exploited by the species pair. Otherwise the model would constitute a perpetuum mobile. By analogy with the competition model this would be avoided by requiring that $\beta_{12}\beta_{21}$ is strictly less than unity. When this is fulfilled the model always guarantees stable coexistence of the two species.

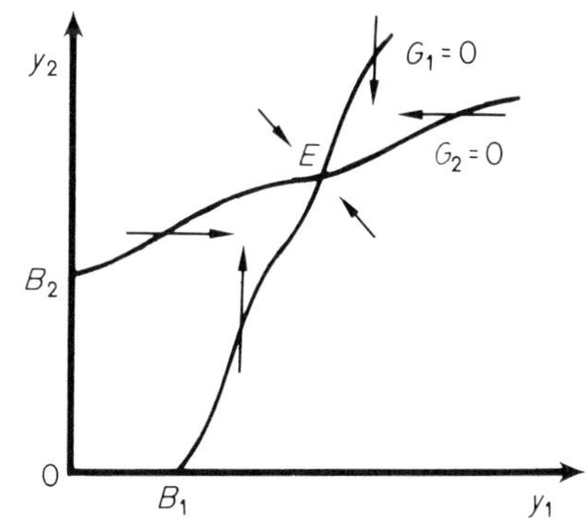


Fig. 30. A general case of two mutualistic species

The model can be formulated in the general framework of model Eq.(2.6) and model Eq.(2.14) by supposing

$$dy_i/dt = y_i G_i(y_1, y_2) \quad \text{for} \quad i=1,2\,, \tag{2.16}$$

with the following restraints on the functions $G_i$.

1. If both populations are small they can multiply, i.e., $G_i(0,0)>0$.

2. Each species when alone cannot multiply over a certain size (the carrying capacity), i.e., $G_1(B_1,0)=G_2(0,B_2)=0$ for $B_i>0$.

3. The two species interact mutualistically meaning that the per capita increase of one species is increased by an increase in the population size of the other species, i.e., $\partial G_i/\partial y_j>0$, $i \neq j$.

4. There exists a ratio $y_1/y_2$ where an increase in the population sizes decreases the per capita growth in each of the species, i.e., $y_1(\partial G_i/\partial y_1)+y_2(\partial G_i/\partial y_2)<-e$, with $e>0$. This condition secures that the system is not a perpetuum mobile.

Under these conditions Eq.(2.16) will always have one globally stable equilibrium of coexistence (Fig. 30). Condition 1 is not realistic for many mutualistic species pairs, i.e., often none of the species can grow without the presence of the other species, so that there is a minimum population size of species $j$ below which $i$ cannot increase. The model (Fig. 30) can also describe such cases. There will then be an unstable equilibrium point $U$ and for population sizes below $U$ both species will go extinct and for population sizes above $U$ the populations will reach the stable equilibrium $E$.

## 2.4 Linear Models of Food Webs

The simple models of prey-predator relationships, competition and mutualism discussed in the previous sections may readily be extended to describe whole food webs. A community with $n$ interacting species may be modelled by equations where the per capita growth rate is a linear function of the population sizes $z_i$. Thus

$$(1/z_i)\,dz_i/dt = w_i + \sum_{j=1}^{n} \omega_{ij} z_j \tag{2.17}$$

where the signs of $\omega_{ij}$ are in accordance with the classification of interrelationships by Odum (1971) as discussed in the introduction to this Chapter, and $w_i$ corresponds to the unlimited growth rate if species $i$ is a primary producer and to the death rate if species $i$ belongs to higher trophic levels. An equilibrium $(\hat{z}_1, \hat{z}_2, \ldots, \hat{z}_n)$ with $\hat{z}_i > 0$ for all $i$, is determined by the set of linear equations

$$-w_i = \sum_{j=1}^{n} \omega_{ij} \hat{z}_j, \qquad i = 1, 2, \ldots, n, \tag{2.18}$$

provided the unique solution is positive ($\hat{z}_i > 0$ for all $i$). The global stability properties of the system [Eq.(2.17)] have not been determined [similar to those of Kolmogoroff (1936) and Rescigno and Richardson (1967)]. We must rely on an analysis of the local stability of the equilibrium as determined by Eq.(2.18). Such mathematical techniques have been developed by May (1973a) and by Strobeck (1973).

A local stability analysis is done by considering a small perturbation from the equilibrium such that the population sizes are given by $z_i = \hat{z}_i + e_i$, where $e_i$ is small. With this change in variables we have

$$de_i/dt = (\hat{z}_i + e_i) \sum_{j=1}^{n} \omega_{ij} e_j \tag{2.19}$$

from Eqs. (2.17) and (2.18). The behavior of the system around the equilibrium is determined by neglecting terms of the order $e^2$ (Gantmacher, 1971), so that Eq. (2.19) becomes the linear system

$$de_i/dt = \sum_{j=1}^{n} (\omega_{ij} \hat{z}_i)\, e_j\,. \tag{2.20}$$

The behavior around the equilibrium is thus determined by the community matrix **A** (Levins, 1968) with the elements

$$a_{ij} = \omega_{ij} \hat{z}_i\,. \tag{2.21}$$

(For general nonlinear systems the same result applies with the element $a_{ij}$ of the community matrix being $\hat{z}_i\{\partial[(1/z_i)dz_i/dt]/\partial z_j\}$. The solution of the system of linear differential Eq. (2.20) may be written

$$e_i = \sum_{j=1}^{n} k_{ij} \exp(\lambda_j t) \tag{2.22}$$

where $\lambda_i$, $i = 1, 2, \ldots, n$, are the eigenvalues of the matrix **A** determined as the real or complex roots of the equation

$$\det(\mathbf{A} - \lambda \mathbf{I}) = 0\,, \tag{2.23}$$

(**I** is the identity matrix) and where $k_{ij}$ are constants determined by the initial conditions (Gantmacher, 1960). The form [Eq. (2.22)] gives a stability criterion. If $\lambda_i > 0$ or the real part $\mathrm{Re}(\lambda_i) > 0$, then the term $\exp(\lambda_i t)$ will increase without limit. On the other hand, if for all $i$, we have that $\mathrm{Re}(\lambda_i)<0$, then the terms will vanish towards zero as time goes on, meaning that the populations return to their equilibrium values. Thus a valid equilibrium is locally stable if the eigenvalues of the community matrix **A** [the roots of Eq. (2.23)] all have negative real parts, and the equilibrium is unstable if only one of the eigenvalues has a positive real part (Gantmacher, 1971).

This criterion of stability is not entirely useful. If the eigenvalues cannot be determined, then the only criterion for negative real parts of all roots of a polynomial equation like Eq. (2.23) is the Routh-Hurwitz criterion (Gantmacher, 1971; May, 1973a; Strobeck, 1973) which consists of purely algebraic conditions with no immediate appeal to the applicant. The criterion, however, is valuable for analyzing simple web models where the number of species is small.

As an example of a local stability analysis we may use the predator–prey model (4) Eq. (2.4). The community matrix for this system is

$$\mathbf{A}=\begin{Bmatrix} a_{11} & a_{12} \\ a_{21} & a_{22} \end{Bmatrix} \sim \begin{Bmatrix} - & - \\ + & - \end{Bmatrix}. \tag{2.24}$$

The polynomial Eq. (2.23) now reads

$$\lambda^2-(a_{11}+a_{22})\,\lambda+(a_{11}a_{22}-a_{12}a_{21})=0\,. \tag{2.25}$$

The criterion for negative real parts of both roots of Eq. (2.25) is that the coefficient to $\lambda$ and the constant term are both positive. An inspection of the signs of the elements of the community matrix [Eq. (2.24)] reveals that this condition is always fulfilled. Thus a model with the signs of Eq. (2.24) is qualitatively stable. On the other hand, if the sign of $a_{11}$ is $+$, as in the model shown on Figure 19, then the criterion for negative real parts of Eq. (2.25) becomes a condition on the actual values of the parameters.

As already discussed in the section on predator–prey systems, a local analysis may have a limited applicability to global reasoning.

## 2.5 Competition Models with Many Species

The linear Lotka-Volterra model for competition described by Eq. (2.12) readily generalizes to more species, i.e.,

$$dx_i/dt=r_ix_i(K_i-x_i-\sum_{j\neq i}\beta_{ij}x_j)/K_i\,. \tag{2.26}$$

For the subsequent analysis it is convenient to change the parameterization of Eq. (2.26) by measuring the population sizes as fractions of the carrying capacities. Eq. (2.26) then reads

$$dx_i/dt=r_ix_i(1-x_i-\sum_{j\neq i}(\beta_{ij}K_j/K_i)\,x_j)\,,$$

or by setting

$$\alpha_{ii}=1 \quad \text{and} \quad \alpha_{ij}=\beta_{ij}K_j/K_i \quad \text{for} \quad i\neq j\,, \tag{2.27}$$

then Eq. (2.26) becomes

$$dx_i/dt=r_ix_i\left(1-\sum_{j=1}^{m}\alpha_{ij}x_j\right). \tag{2.28}$$

An equilibrium $(\hat{x}_1, \hat{x}_2, \ldots, \hat{x}_m)$ with all $m$ species present will be the solution to the linear equations

$$1 = \sum_{j=1}^{m} \alpha_{ij}\hat{x}_j, \qquad i = 1, 2, \ldots, m, \tag{2.29}$$

and the solution is unique (except for pathological cases) and a valid equilibrium is found when $\hat{x}_i > 0$ for all $i$.

The stability analysis of this equilibrium is more tricky, however, because the simple graphical arguments used for the two-species model cannot be used here. A general global analysis of the model is not known, so we will have to rely on a local stability analysis of the equilibrium $(\hat{x}_1, \hat{x}_2, \ldots, \hat{x}_m)$. This analysis is outlined by May (1973a) and by Strobeck (1973), but the stability conditions were given in terms of the rather opaque Routh-Hurwitz criterion which may be used in specific models but does not provide general insight in the stability properties. Strobeck (1973) considered the three species version of Eq. (2.28) in more detail in order to investigate the connection between the stability criterion and the condition that the solution to Eq. (2.29) is a valid equilibrium. He showed that the condition for the increase of a third species invading a previously stable two species equilibrium implies a part of the stability condition for the three species equilibrium. Eventually, if all two species marginal equilibria are stable in the absence of the third species, and the missing species can invade these equilibria, then the internal three species equilibrium exists and is stable.

A special type of competition model may be analyzed in much more detail, viz., models with symmetric competition where either $\alpha_{ij} = \alpha_{ji}$ or $\beta_{ij} = \beta_{ji}$ for all $i$ and $j$. We will here analyze the first possibility. First observe that the function

$$Q(t) = \sum_{i=1}^{m} \sum_{j=1}^{m} (x_i - \hat{x}_i)\, \alpha_{ij}(x_j - \hat{x}_j) \tag{2.30}$$

acts like an energy function (Lyapunov function) of the system, i.e., $dQ/dt$ is always negative, see Eq. (2.32) (MacArthur, 1970; May, 1974).

To see this differentiate to get

$$dQ/dt = \sum_{i=1}^{m} \sum_{j=1}^{m} \alpha_{ij}\{(x_i - \hat{x}_i)\, dx_j/dt + (x_j - \hat{x}_j)\, dx_i/dt\} = 2 \sum_{i=1}^{m} \sum_{j=1}^{m} \alpha_{ij}(x_j - \hat{x}_j)\, dx_i/dt \tag{2.31}$$

due to the symmetry of the two addents. From Eq. (2.28) and the equilibrium Eq. (2.29) we get

$$dx_i/dt = -r_i x_i \sum_{k=1}^{m} \alpha_{ik}(x_k - \hat{x}_k)\,.$$

Combining this with Eq. (2.31) reveals that

$$dQ/dt = -2 \sum_{i=1}^{m} r_i x_i \left[ \sum_{j=1}^{m} \sum_{k=1}^{m} \alpha_{ij}\alpha_{ik}(x_j - \hat{x}_j)\,(x_k - \hat{x}_k) \right]$$

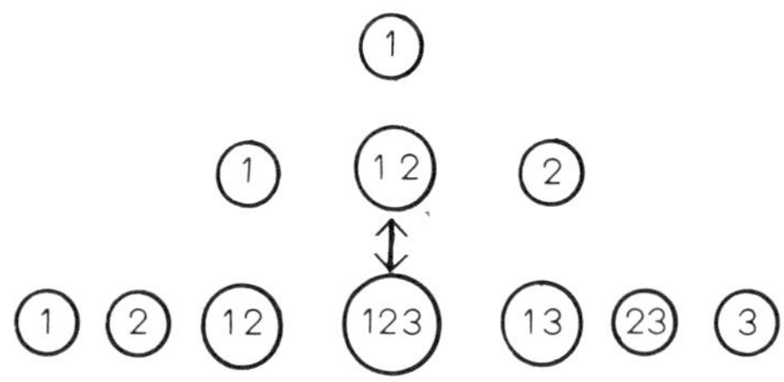


Fig. 31. Alternative community compositions of systems consisting of 1, 2, or 3 competing species respectively

which simplifies to

$$dQ/dt = -2 \sum_{i=1}^{m} r_i x_i \left[ \sum_{j=1}^{m} \alpha_{ij}(x_j - \hat{x}_j) \right]^2 . \qquad (2.32)$$

The population sizes are nonnegative so Eq. (2.32) is nonpositive under the simple condition that $r_i > 0$ for all $i$. Therefore, the population will grow such that $Q$ decreases and a stable equilibrium will be a local minimum of $Q(x)$. At the equilibrium $(\hat{x}_1, \hat{x}_2, \ldots, \hat{x}_m)$ we see that $Q = 0$, i.e., the equilibrium is stable if $Q > 0$ in the neighborhood around the equilibrium. However, $Q$ is a quadratic form over the matrix $\mathbf{B} = (\alpha_{ij})$, so the requirement that $Q > 0$ in the neighborhood of the equilibrium is that $Q$ is positive definite, i.e., positive everywhere except at the equilibrium. The matrix $\mathbf{B}$ is symmetric so its eigenvalues are real (cf. Gantmacher, 1960) meaning that $Q$ is positive definite provided all the eigenvalues of $\mathbf{B}$ are positive. A valid equilibrium determined by Eq. (2.29) is, therefore, globally stable provided all the eigenvalues of the symmetric matrix $\mathbf{B}$ are positive (MacArthur, 1970; Strobeck, 1973).

The influence of the number of species on the stability of an equilibrium may be inferred from the following properties of symmetric matrices (MacArthur, 1971; cf. also Gantmacher, 1960). If we remove the last row and the last column from a symmetric matrix then the least eigenvalue is greater than or equal to that of the original matrix. Now, the criterion for stability being that all eigenvalues of the symmetric matrix $\mathbf{B}$ should be positive (the least eigenvalue positive) shows that if we have stable coexistence of $m$ species then we have stable coexistence for any lower number of species for which a valid equilibrium exists.

In symmetric models this may be illustrated by a simple example as shown in Figure 31, where we construct a community of an increasing number of competing species. Beginning with species 1 and then introducing species 2 we get two possibilities for a change in the community composition, i.e., that species 2 excludes species 1 or that the two species coexist. However, the coexistence of the competitors imply that species 2 can maintain a population in the habitat so that the stability of the two species system implies the stability of either of the two monospecies equilibria if one of the species is removed. By introducing species 3, there are 4 new possibilities of species composition. Again the stability of the multispecies equilibrium (123) implies the marginal stability of the two and one species equilibria. Thus, the possibility of going to a higher level of complexity is limited as compared going to equivalent complexities or to going down in species numbers.

This sequence is most easily seen in the simple model

$$\mathbf{B}=\begin{Bmatrix} 1 & \alpha & \alpha^* \\ \alpha & 1 & \alpha \\ \alpha^* & \alpha & 1 \end{Bmatrix} \tag{2.33}$$

as discussed in the following section of this chapter. Deviations from the general behavior of symmetric models may be demonstrated by the model

$$\mathbf{B}=\begin{Bmatrix} 1 & \alpha & \alpha^* \\ \alpha^* & 1 & \alpha \\ \alpha & \alpha^* & 1 \end{Bmatrix} \tag{2.34}$$

which has the same properties as the symmetric model [Eq. (2.33)] as long as $\alpha$ and $\alpha^*$ are both $<1$, i.e., as long as all two species coexistence equilibria exist and are stable. However, when $\alpha^*>1>\alpha$, then neither of the two species equilibria exist but a three species equilibrium with all species being equally abundant exists. In this situation all the one species equilibria are unstable in the three species system, so all boundaries repel the population. However, the three species equilibrium may only be stable for $1>\alpha\alpha^*$; when this condition is not fulfilled, none of the equilibria are stable and the behavior is necessarily cyclical. Note, however, that we have not demonstrated the existence of a stable limit cycle.

Mutualistic systems which are symmetric share the property with the symmetric, competitive systems that the quadratic form is a global Lyapunov function. Thus the remarks on the influence of species numbers on the stability of symmetric systems applies to mutualistic systems too. However, the inherent two species condition for the validity of the model, $\alpha_{ij}\alpha_{ji}<1$, or for three species that $\det(\mathbf{B})>0$, which guarantees that the system does not grow unlimited, is also the condition for $Q$ being positive definite (cf. Gantmacher, 1960) in the three species model. Thus, mutualistic systems are stable whenever they are limited in growth.

## 2.6 Simple Food Web Models

Predator-prey interactions are in general stable as discussed previously in this chapter whereas competing species within a trophic level lack this kind of inherent stability. In order to get a feeling for the influence of predation on the interaction between competing species consider the simple food web on Figure 32 and let the species $x_1$, $x_2$, and $x_3$ compete in a symmetric fashion with the competition matrix [Eq. (2.33)], and let the predator, $y$, be nondiscriminative such that

$$\begin{aligned} dx_1/dt &= rx_1(1-x_1-\alpha x_2-\alpha^* x_3-\gamma y)\,, \\ dx_2/dt &= rx_2(1-\alpha x_1-x_2-\alpha x_3-\gamma y)\,, \\ dx_3/dt &= rx_3(1-\alpha^* x_1-\alpha x_2-x_3-\gamma y)\,, \\ dy/dt &= y(-d+\gamma x_1+\gamma x_2+\gamma x_3-\delta y)\,, \end{aligned} \tag{2.35}$$

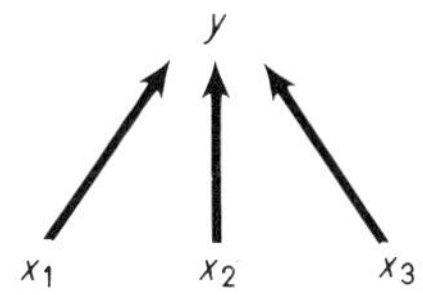


Fig. 32. A community consisting of 3 competing species and a predator

Let us first consider the system with $y=0$ and analyze the sequence of invasions such as is shown on Figure 31. With one species the equilibrium is $\hat{x}_1=1$. If species 3 is introduced it will increase provided that $\alpha^*<1$ and the populations will then go to the stable equilibrium

$$\hat{x}_1=\hat{x}_3=1/(1+\alpha^*)\,. \tag{2.36}$$

If we now introduce species 2 it will increase if

$$1-\alpha(\hat{x}_1+\hat{x}_3)>0\,, \tag{2.37}$$

or, using Eq. (2.36) we have the condition as

$$(1+\alpha^*)/2>\alpha\,. \tag{2.38}$$

If Eq. (2.38) is satisfied we will have the stable equilibrium

$$\begin{aligned}\hat{x}_1&=\hat{x}_3=(1-\alpha)/(1+\alpha^*-2\alpha^2)\,,\\ \hat{x}_2&=(1+\alpha^*-2\alpha)/(1+\alpha^*-2\alpha^2)\,.\end{aligned} \tag{2.39}$$

The influence of the predator on the conditions for invasions and coexistence of the competitors may now be analyzed by going through the same sequence of events in the presence of $y$. In the absence of $x_2$ and $x_3$ the equilibrium of Eq. (2.35) is

$$\begin{aligned}\hat{x}_1&=(\delta+d\gamma)/(\delta+\gamma^2)\,,\\ \hat{y}&=(\gamma-d)/(\delta+\gamma^2)\end{aligned} \tag{2.40}$$

which is valid for $\gamma>d$. This is also the condition for global stability according to our previous results in section 2.1.

Now introduce species 3 and observe from Eq. (2.35) that it will increase if

$$1-\alpha^*\hat{x}_1-\gamma\hat{y}>0\,. \tag{2.41}$$

Inserting Eq. (2.40) into Eq. (2.41) gives the condition that $\alpha^*<1$. Thus the condition for the invasion of species 3 is independent of the presence of the predator.

The equilibrium with the prey species 1 and 3 and the predator is

$$\hat{x}_1 = \hat{x}_3 = (\delta + \gamma d)/[\delta(1+\alpha^*) + 2\gamma^2],$$
$$\hat{y} = (2\gamma - d(1+\alpha^*))/[\delta(1+\alpha^*) + 2\gamma^2], \tag{2.42}$$

which is valid for $2\gamma > d(1+\alpha^*)$ and locally stable when $\alpha^* < 1$. The condition for the validity of the equilibrium [Eq. (2.42)] is less restrictive than the condition for the validity of the equilibrium [Eq. (2.40)]. Thus, if the one-predator-one-prey system is stable so is the one-predator–two-prey system, if the two prey species can coexist. This conclusion is not surprising since an extra prey species is just a resource supply for the predator; a less efficient predator might not be able to maintain a population with one of the prey species but might be able to increase in the presence of two prey species. In terms of Figure 18 this amounts to the point $b$ being above or below the point $A$.

Finally, let us introduce species 2. From Eq. (2.35) we observe that it will increase if

$$1 - \alpha(\hat{x}_1 + \hat{x}_3) - \gamma\hat{y} > 0. \tag{2.43}$$

Inserting Eq. (2.42) into Eq. (2.43) reveals that condition [Eq. (2.43)] is identical with condition [Eq. (2.38)]. Again the presence of the predator does not change the qualitative outcome of the competition.

This small story shows very nicely that the stabilizing effect of predation on a set of competing species may be very limited. However, we have to keep in mind that the model used here has a restricted value due to its pronounced symmetries.

A system which is not symmetric with respect to competition is

$$dx_1/dt = r_1 x_1(1 - x_1 - \alpha_{12} x_2) - \bar{r}\gamma x_1 y,$$
$$dx_2/dt = r_2 x_2(1 - \alpha_{21} x_1 - x_2) - \bar{r}\gamma x_2 y, \tag{2.44}$$
$$dy/dt = y(-d + \gamma x_1 + \gamma x_2 - \delta y),$$

where $\bar{r} = (r_1 + r_2)/2$. By setting the derivatives [Eq. (2.44)] to zero we find the equilibrium as

$$\hat{x}_1 = [r_1 r_2(1-\alpha_{12})\,\delta + \bar{r}(r_1 - r_2)\,\gamma^2 + \bar{r}(r_2 - \alpha_{12} r_1)\,\gamma d]/D, \tag{2.45}$$

$$\hat{x}_2 = [r_1 r_2(1-\alpha_{21})\,\delta + \bar{r}(r_2 - r_1)\,\gamma^2 + \bar{r}(r_1 - \alpha_{21} r_2)\,\gamma d]/D, \tag{2.46}$$

$$\hat{y} = \{r_1 r_2[(2 - \alpha_{12} - \alpha_{21})\,\gamma - (1 - \alpha_{12}\alpha_{21})\,d]\}/D, \tag{2.47}$$

where

$$D = r_1 r_2(1 - \alpha_{12}\alpha_{21})\,\delta + \bar{r} r_1(1-\alpha_{12})\,\gamma^2 + \bar{r} r_2(1-\alpha_{21})\,\gamma^2. \tag{2.48}$$

If $x_2 = 0$ then we have the equilibrium

$$\hat{x}_1 = (r_1\delta + \bar{r}\gamma d)/(r_1\delta + \bar{r}\gamma^2),$$
$$\hat{y} = r_1(\gamma - d)/(r_1\delta + \bar{r}\gamma^2), \tag{2.49}$$

which is valid and stable for $\gamma > d$ [cf. Eq. (2.40)]. Now $x_2$ can invade if the numerator of Eq. (2.46) is positive. Similarly, if $x_1 = 0$, it can invade if the numerator of Eq. (2.45) is positive. It can now be seen from Eqs. (2.45) and (2.46) that when $\alpha_{12}$ and $\alpha_{21} < 1$ then $x_1$, $x_2$ and $y$ can only coexist in a certain interval, $a < r_1/r_2 < b$ with $a < 1 < b$, i.e., we have in this case a stricter criterion for coexistence than if the predator is absent. For $\delta$ small the condition for coexistence is approximately

$$(\gamma - d)/(\gamma - \alpha_{12}d) < r_1/r_2 < (\gamma - \alpha_{21}d)/(\gamma - d). \tag{2.50}$$

If $\alpha_{12} < 1$ and $\alpha_{21} > 1$ with $\alpha_{12}\alpha_{21} < 1$ it can be argued that coexistence of all three species may be possible. First, the condition $\gamma > d$ has to be replaced by the stronger condition $\hat{y} > 0$. The conditions for positivity of $\hat{x}_1$ and $\hat{x}_2$ for $\delta$ small are again approximately Eq. (2.50), and the condition that this interval is nonempty is equivalent to the condition $\hat{y} > 0$, which also insures that $D > 0$ in the interval Eq. (2.50). The right term in condition (2.50) is in this case less than unity, so that coexistence is only possible if $r_2 > r_1$, i.e., if the species that suffers from competition can compensate for this by a larger reproductive potential.

This example again shows that it cannot be taken as a generality that predation makes coexistence of competitors possible when it is impossible in the absence of the predator. In the model [Eq. (2.44)] we found that the criterion for coexistence in the absence of the predator (i.e., that both $\alpha$'s are less than unity) is insufficient in the presence of the predator. Only for a certain range of the ratio $r_1/r_2$ can we have coexistence. However, in a similar interval we may have coexistence when one of the competition coefficients exceeds one. Thus, in this model, the predator is in general restricting the interval of coexistence, but is able to move it around.

More elaborate predator-prey models which take learning by the predator or other behavioral characteristics into account may predict a stabilizing effect on the community of prey species. Roughgarden and Feldman (1975) considered a model in which the predation pressure on rare species is lower than on common species due to a habituation to the most frequent prey species by the predator. This effect enhances the coexistence of more competing prey species.

In spite of the general conclusion we arrived at above (i.e., that predation does not as a general principle stabilize the coexistence of competing prey species) several authors (e.g., Slobodkin, 1961, 1964; Paine, 1966) have, based on verbal theory or empirical evidence, arrived at the conclusion that predation enhances the diversity of biotic communities. We will here discuss some of these examples including observational evidence which exemplify the model of Roughgarden and Feldman (1975).

Some laboratory systems which are described in the literature seem to exemplify a stabilizing effect of predators on competing prey species, but not all these experiments have been analyzed. For example, Utida (1953) enhanced the coexistence of two species of bean weevils by adding to the experimental systems a parasitoid wasp which attacks both species.

Slobodkin (1964) studied the competition between two species of *Hydra* in laboratory cultures. In these systems one species always excludes the other. When a constant fraction of both species were removed regularly, however, coexistence or at least a very long persistence of both species was obtained. In these experiments unselective predation was apparently simulated. However, in this experiment the predator (i.e., the scientist) did not respond to changes in the prey

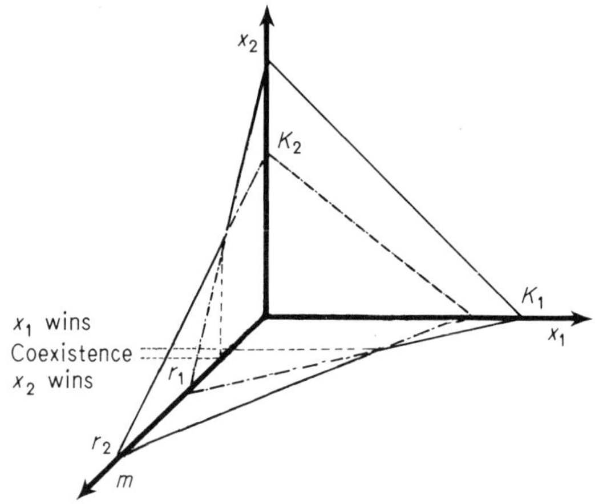


Fig. 33. A graphical analysis of a system of two competing species in which a constant fraction, $m$, is removed from both populations per time unit. (Redrawn from Slobodkin, 1961)

population by increasing in numbers or by starving, in contrast to the models considered above. A model describing this experiment was considered by Slobodkin (1961). Since a constant fraction of both prey species are removed per time unit we can write;

$$dx_1/dt = x_1 r_1 (K_1 - x_1 - x_2\beta_{12})/K_1 - mx_1\,,$$
$$dx_2/dt = x_2 r_2 (K_2 - x_2 - x_1\beta_{21})/K_2 - mx_2\,, \tag{2.51}$$

where $m$ is a constant removal factor. Consider a system in which, say $x_1$ always wins independently of initial conditions when $m=0$, i.e., $K_1/\beta_{12} > K_2$ and $K_1 > K_2/\beta_{21}$ (see Section 2.2) and further let $r_2 > r_1$. This system can easily be analyzed graphically like the Lotka-Volterra model for competition (Fig. 27) by letting $m$ be represented by a third axis perpendicular to the $(x_1, x_2)$-plane (Fig. 33). It is seen that the species with the highest value of $r$ is favored by the "predation" (obviously, when $r_1 < m < r_2$ then species 1 will go extinct while species 2 will persist) and that in a certain interval of values of $m$ the two prey species coexist. The effect of imposing a removal rate $m$ on the system corresponds to changing the carrying capacity, $K_i$, to $K_i(1 - m/r_i)$ [see Eq. (1.8)]. This model may be relevant for predator-prey systems where the predator has a very much longer generation time than the prey species.

As mentioned above, the habituation of a predator to the prey species which is the most abundant one at any time may stabilize the coexistence of competing prey species. There is evidence to show that prey switching is not uncommon. Fisher-Piette (1934) showed that the intertidal snail *Nucella lapillus* preferably feeds on barnacles when these are abundant but when mussels dominate, the latter are preferred. Lawton et al. (1974) studied the prey selectivity of the aquatic hemipteran *Notonecta*. The insects were kept in containers with two

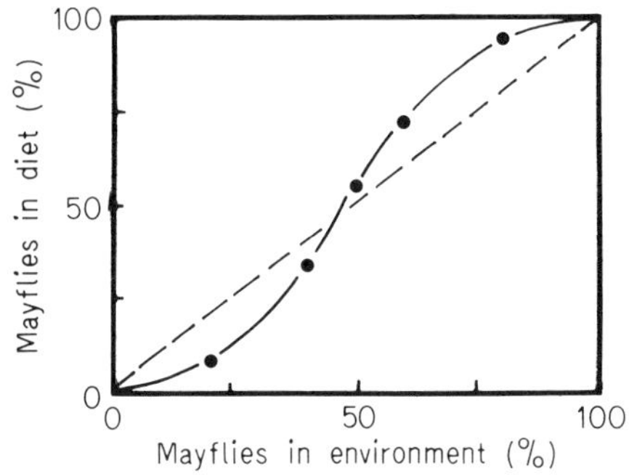


Fig. 34

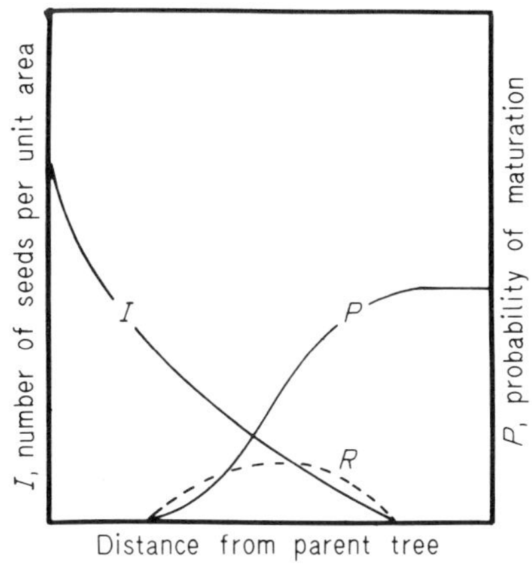


Fig. 35

Fig. 34. The percentage of mayflies in the diet of *Notonecta* as a function of the percentage of mayflies of the potential prey species. (Redrawn from Lawton et al., 1974)

Fig. 35. The recruitment ($R$) of new trees in rainforests as function of the distance of the parent tree. (Redrawn from Janzen, 1970)

prey species, the isopod *Asellus* and the mayfly *Clöeon*, which were added in different proportions. After 8–10 days *Notonecta* preyed supraproportionally on the most frequent prey species (Fig. 34). Lawton et al. (1974) also showed that *Notonecta* became more successful in attacking *Asellus* after having been kept in containers with a high density of this isopod for some time. Thus a learning process probably plays a role for *Notonecta*. Other examples on switching in predators are given by Murdoch (1969).

Tropical rainforests are known to be rich in species of trees; more than 100 species may be found within one hectare. Janzen (1970) explains this as a result of seed herbivory. The underlying mechanism is that seeds falling immediately below the mother tree have a very small chance of germinating due to prey specific and density responsive insects. These latter gather where their preferred seed species are most numerous whereas single, isolated seeds which have become more dispersed from the mother tree have a higher probability of germination and maturation (Fig. 35). This mechanism will allow for a higher number of coexisting trees since it prevents continuous stands of any one species. Janzen (1970) also gives evidence that the assumptions of his model, i.e., specificity and density responsiveness of tropical seed predators, hold true.

Paine's (1966) work on species richness on intertidal rocks is probably the most well known example of the effect of a predator on the coexistence of its prey species. This study, however, is more appropriately discussed together with migration-extinction equilibria in Chapter 4.

# 3. The Niche and Limiting Similarity

## 3.1 The Concept of the Niche

Chapter 2 shows in a general way that there is a limit to the similarity of coexisting species with respect to the sharing of limiting resources. If we consider qualitatively different categories of resources, a simple graphical model described by MacArthur and Levins (1964) shows that the number of coexisting species can at most equal the number of limiting resources. Consider a $(\mathscr{R}_1, \mathscr{R}_2)$ plane where $\mathscr{R}_1$ and $\mathscr{R}_2$ represent the quantities of two different kinds of renewable resources. Further, assume that there are two species, $x_1$ and $x_2$, which both utilize both kinds of resources but with different efficiencies. We might assume that

$$
\begin{aligned}
dx_1/dt &= x_1(i_{11}\mathscr{R}_1 + i_{12}\mathscr{R}_2 - \mathscr{T}_1)\,, \\
dx_2/dt &= x_2(i_{21}\mathscr{R}_1 + i_{22}\mathscr{R}_2 - \mathscr{T}_2)\,,
\end{aligned}
\tag{3.1}
$$

where $i_{jk}$ is the efficiency with which species $j$ utilizes resource $k$ and $\mathscr{T}_j$ is the threshold value for the combined availability of the two resources which allow species $j$ to grow. In the $(\mathscr{R}_1, \mathscr{R}_2)$ plane we can draw the lines $dx_j/dt=0$ according to Eq. (3.1). These lines represent the resource combinations which can just sustain stable populations of $x_1$ and $x_2$ (Fig. 36). [Note: This diagram is different from the $(x_1, x_2)$ planes used for analysis of competition models in the previous chapter. In Figure 36 the populations show positive growth when the resource combination corresponds to a point *outside* the $dx_j/dt=0$ lines.]

It can be seen from Figure 36 that if each of the species utilizes one of the resources more efficiently, i.e., the two lines $dx_j/dt=0$ intersect in the positive quadrant, then the two species can seemingly coexist in a resource combination corresponding to the point of intersection. It can also be seen that if a third species is introduced it will either not be able to establish a population or it will exclude one of the species initially present (unless the three lines intersect in one point which can be considered as extremely unlikely). The arguments can easily be extended to a system with three types of resources which will then at most be able to sustain stable populations of three competing species.

The model presupposes that the coexisting species can distinguish between the different kinds of resources. If the two species cannot distinguish between $\mathscr{R}_1$ and $\mathscr{R}_2$ but eat them in the proportions in which they happen to occur, then the two lines, $dx_j/dt=0$, are parallel and the two species will not be able to coexist.

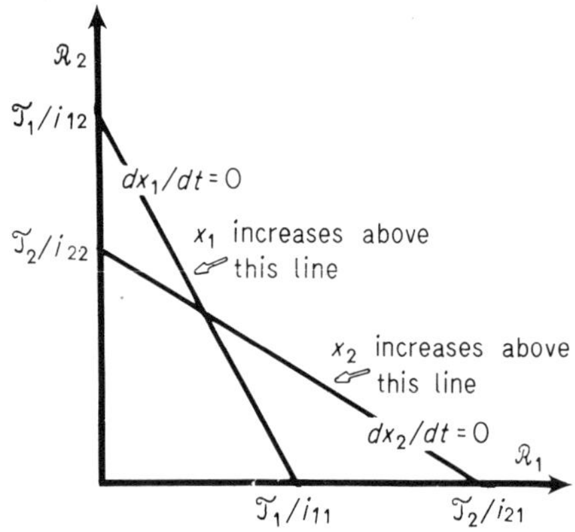


Fig. 36. A graphical representation of the competition model [Eq. (3.1)]

In fact, there is in this case only one kind of resource, namely the combined resource measured by $\mathscr{R}_1+\mathscr{R}_2$. MacArthur and Levins (1964) termed this kind of resource utilization *fine grained*. The resource "grains" are too small for the species to distinguish between the different qualities. On the other hand, a utilization which distinguishes the resource qualities is called *coarse grained*. A filterfeeding animal which utilizes a wide variety of microplankton species in the proportion they occur in the water utilizes its resource in a fine grained manner, relative to say a copepod which can specialize on, e.g., diatoms of a certain size range. Other terms for this distinction are generalist and specialist species.

As the model has been presented so far, there has been no specific assumption about the availability of $\mathscr{R}_1$ and $\mathscr{R}_2$ as a function of the use, and if only $i_{11}/i_{21}$ and $i_{12}/i_{22}$ are both greater than $\mathscr{T}_1/\mathscr{T}_2$ then coexistence is predicted. We now assume that the resources $\mathscr{R}_1$ and $\mathscr{R}_2$ grow in such a way that the balance between growth and utilization at any time provides the resource abundance

$$\mathscr{R}_k=\mathscr{S}_k-c_1 i_{1k} x_1-c_2 i_{2k} x_2 \quad \text{for} \quad k=1,2, \tag{3.2}$$

where $\mathscr{S}_k$ is the abundance of resource $k$ when not exploited and $c_j$ is a constant measuring the conversion efficiency of the resources into individuals of species $j$. Letting $c_j=1$ for simplicity (corresponding to measuring the population sizes of the two species in units of their weight on the resources), then Eq. (3.2) for the resources allows Eq. (3.1) to be expressed in terms of the population sizes only:

$$\begin{aligned} dx_1/dt&=x_1[(i_{11}\mathscr{S}_1+i_{12}\mathscr{S}_2)-(i_{11}^2+i_{12}^2)x_1-(i_{11}i_{21}+i_{12}i_{22})x_2-\mathscr{T}_1],\\ dx_2/dt&=x_2[(i_{21}\mathscr{S}_1+i_{22}\mathscr{S}_2)-(i_{21}^2+i_{22}^2)x_2-(i_{11}i_{21}+i_{12}i_{22})x_1-\mathscr{T}_2], \end{aligned} \tag{3.3}$$

or it may be written as the ordinary competition equations used in the last chapter:

$$\begin{aligned} dx_1/dt&=x_1 r_1(K_1-x_1-\beta_{12}x_2)/K_1,\\ dx_2/dt&=x_2 r_2(K_2-x_2-\beta_{21}x_1)/K_2, \end{aligned} \tag{3.4}$$

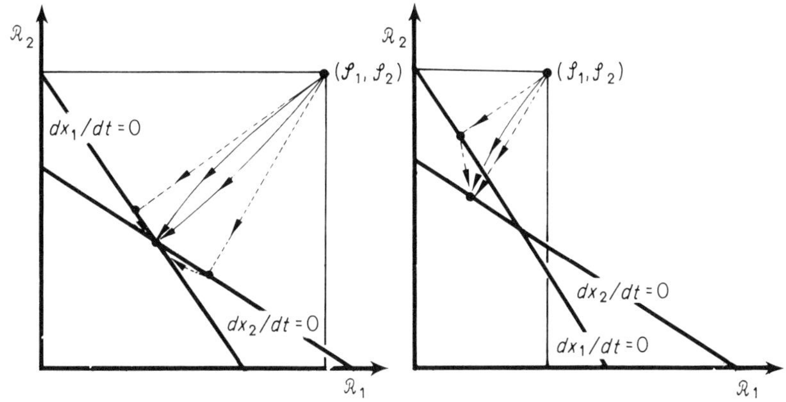


Fig. 37. A graphical representation of the competition model [Eqs. (3.4) and (3.5)] where the carrying capacities ($S_1$ and $S_2$) of the two kinds of resources are taken into consideration. Broken arrows: the resource trajectories when only one of the species is present. *Left:* both species will persist; *right*: species 1 will be excluded

where these parameters now are expressed by

$$\begin{aligned} r_j &= (i_{j1}\mathscr{S}_1 + i_{j2}\mathscr{S}_2 - \mathscr{T}_j) \\ K_j &= (i_{j1}\mathscr{S}_1 + i_{j2}\mathscr{S}_2 - \mathscr{T}_j)/(i_{j1}^2 + i_{j2}^2)\,. \\ \beta_{jj'} &= (i_{11}i_{21} + i_{12}i_{22})/(i_{j1}^2 + i_{j2}^2)\,. \end{aligned} \tag{3.5}$$

Equation (3.1) thus has the form of the Lotka-Volterra equations [Eq.(2.12)] and the condition for stable coexistence of the two species is now given by the conditions in Section 2.2,

$$1/K_1 > \beta_{21}/K_2 \quad \text{and} \quad 1/K_2 > \beta_{12}/K_1\,. \tag{3.6}$$

These are now conditions on the resource parameters $(\mathscr{S}_1, \mathscr{S}_2)$, if we consider the parameters $i_{jk}(j, k=1, 2)$ and $\mathscr{T}_j(j=1, 2)$ as given species characteristic parameters. This can be illustrated by the following graphical argument in the $(\mathscr{R}_1, \mathscr{R}_2)$ plane (Fig. 37).

If initially $x_1 = x_2 = 0$, then the resource combination is $(\mathscr{S}_1, \mathscr{S}_2)$. If now $x_1$ is introduced and $x_2$ is absent, then the resource combinations will follow a linear trajectory perpendicular to the line $dx_1/dt=0$ until it reaches this line, since from Eq. (3.2) we have $(\mathscr{R}_1, \mathscr{R}_2) = (\mathscr{S}_1, \mathscr{S}_2) - (i_{11}, i_{12})x_1$ and $dx_1/dt=0$ is given by the equation $i_{11}\mathscr{R}_1 + i_{12}\mathscr{R}_2 - \mathscr{T}_1 = 0$. Similarly, if we keep $x_1=0$ and introduce $x_2$ then the trajectory of $(\mathscr{R}_1, \mathscr{R}_2)$ will be a line perpendicular to the line $dx_2/dt=0$. Now if both species are introduced then the resource combinations must move between these lines. The intersection between the lines $dx_1/dt=0$ and $dx_2/dt=0$ is the only point in the $(\mathscr{R}_1, \mathscr{R}_2)$ space where the two species can have stable coexistence. If this point lies outside the sector of possible resource combinations between the lines through $(\mathscr{S}_1, \mathscr{S}_2)$ perpendicular to the lines $dx_j/dt=0$, then the

$(\mathscr{R}_1, \mathscr{R}_2)$ trajectory cannot reach the equilibrium point and one of the species is excluded (Fig. 37).

From this and Eqs. (3.5) and (3.6) we see that the availability of the resources $\mathscr{R}_1$ and $\mathscr{R}_2$ determines whether the two species can coexist. We can also see from Figure 37 that the more similar the two species are with respect to the use of the resources (i.e., for smaller angle between vectors perpendicular to the lines $dx_j/dt = 0$) the more demanding are the conditions for coexistence on the initial availability $(\mathscr{S}_1, \mathscr{S}_2)$ of the resources. Further, the model provides an interpretation of the competition coefficients in terms of the resource utilization parameters.

The model [Eq. (3.2)] corresponds to logistic growth of the resources with very high growth rate (MacArthur, 1970) which may be seen by using Tikhonov's (1952) approximate method of solving coupled differential equations (cf. Göbber and Seelig, 1975). Thus the model applies for prey-predator systems where the generation time of the prey species is very short compared to that of the predator.

The MacArthur-Levins model described above is in many ways a useful tool for analyzing coexistence and data on, e.g., food selectivity of coexisting congeners. However, organisms may often differ with respect to the utilization of resources which are present in a continuous spectrum of qualities, and organisms may also differ along several such resource dimensions. The model cannot describe such cases. In order to analyze how organisms subdivide their environment and to find limits to the similarity of coexisting species, ecologists have for more than half a century attempted to formalize the concept of the *ecological niche.*

The word "niche" which in different Indoeuropean languages has various meanings, such as nest, corner, hole, dwelling place, and recess, was first used by Grinnell (1917a, b) as an ecological concept. He reported on the "Niche-Relationships of the Californian Thrasher" and he tried to explain the bird's restricted geographical distribution by "adjustment of the bird in various physiological and psychological respects to a narrow range of environmental conditions." To Grinnell, the niche was the "various circumstances" in which an animal lives and he made the distinction to the habitat, i.e., the area inhabited by the creature in question. To Elton (1933) the niche "means the mode of life and especially the mode of feeding of an animal. It is used in ecology in the same sense that we speak of trades and professions in a human community."

Thus, while the verbally formulated concept of the niche goes many years back, it was Hutchinson (1957) who first gave a formalized definition of the niche: "The fundamental niche may be regarded as a set of points in an abstract $n$-dimensional space." The $n$ dimensions are environmental parameters and the points in the fundamental niche corresponds to states "of the environment which would permit the species ... to exist indefinitely." This was the first model which could be manipulated mathematically and used to explain competitive displacement. The *fundamental niche* is defined by all the points in the niche space in which the species can sustain a population; the *realized niche* is defined as the part of the fundamental niche in which the species is not excluded by interspecific competition. In terms of Hutchinson's niche concept, the "competitive exclusion principle" may be formulated in the following way: In a physical space which corresponds to a point in the niche space belonging to the fundamental niche of two species, only one of the species will persist.

Hutchinson's concept of the niche has had a great influence on ecology, theoretically as well as on the observational and experimental field. Yet, it has a number of shortcomings which are best illustrated by the formulation of the competitive exclusion principle; it states that only one species can be present in a physical space corresponding to a point in the niche space, but if this point is in the fundamental niche of any other species, then the theory has no way of predicting which species will be present. The model ignores the fact that there are optimal and suboptimal conditions for growth of a population; in the Hutchinson model the response of the organism is an "all or none." The concept does not take into account that many niche dimensions are correlated and many cannot be arranged linearly as required by the model, e.g., as temporal differences between species. This, however, is a minor point, which only requires that the niche space should be taken more generally than the linear space considered by Hutchinson.

The Hutchinson niche concept has been developed further by MacArthur and Levins (1967) and by Levins (1968) and we will use their formulation of the niche concept in the following. To Levins (1968) the niche is a "fitness measure on an environmental hyperspace.... Thus, corresponding to each point in the environmental hyperspace, there is a measure of the probability of survival and reproduction in that environment." The fundamental niche is the part of the environment (or biotic space) which an organism can exploit according to its phenotype and the realized niche is as before the environments to which the organism is limited by competition. This niche concept is more dynamic, the "realized niche can change rather quickly as a result of environmental change, demographic change, or individual learning. The fundamental niche is modifiable by way of natural selection."

This concept of the niche has been further developed and used for theoretical work by, e.g., MacArthur (1972), Roughgarden (1972, 1974a, b, 1975), May (1973a), and Pianka (1974), and for the analysis of field data by, e.g., Schoener (1968), Pianka (1969, 1973), MacArthur (1972), Roughgarden (1974a, b), and Fenchel (1975b). A thorough discussion of the niche concept is given by Stern and Roche (1974).

## 3.2 Niche Dimensions

The niche has three main dimensions: time, habitat and resource. Organisms which differ in their time niches may live in the same habitat and utilize the same resources but their periods of activity, presence or breeding are displaced; e.g., one species may be day active and the other nocturnal. Many juvenile marine animals migrate into shallow water and estuaries in the summer and emigrate to deeper water in the winter. These juvenile fish and crustaceans feed mainly on the benthic meiofauna (nematodes, small crustaceans, etc.). Figure 38 shows the sequence of arrival and growth through time of the most important animals in a Danish estuary. It can be seen that the species only overlap slightly in size at any

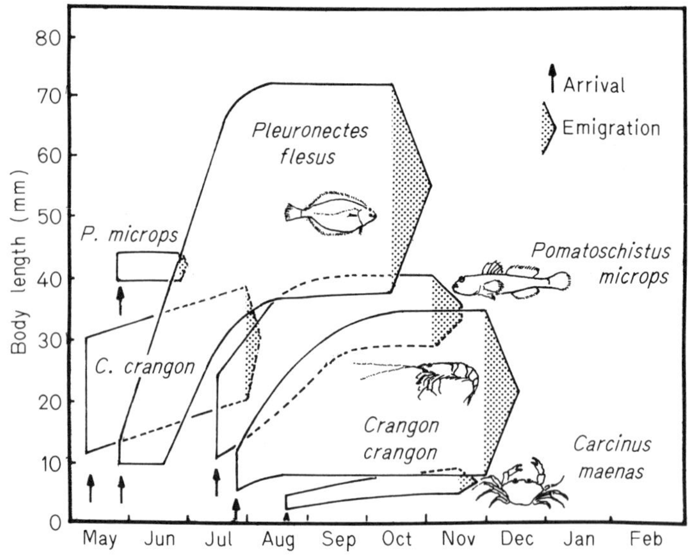


Fig. 38. The arrival, growth and emigration of the vagile macrofauna in a Danish estuary. (Redrawn from Muus, 1967)

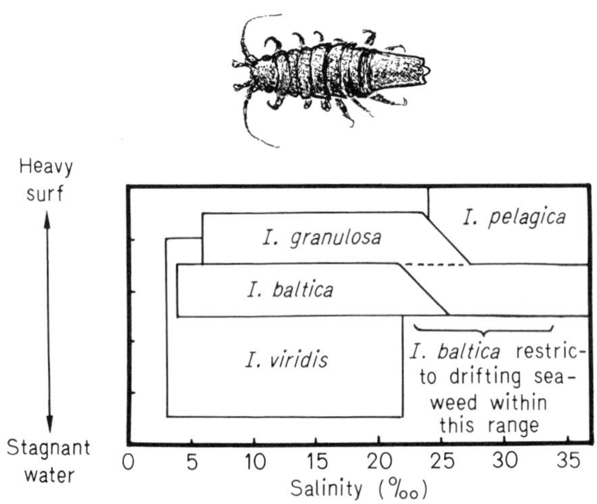


Fig. 39. The distribution of 4 species of the isopod genus *Idothea* in Danish waters according to water turbulence and to salinity. (Redrawn from Muus, 1967)

one time and, therefore, probably only overlap slightly with respect to food size. Thus, these animals have different time niches.

Creatures with different habitat niches may be identical with respect to the utilization of resources and activity periods but differ with respect to habitat preference or have different competitive abilities in different habitats. Factors such as temperature, feeding and perching places, salinity, humidity and substrate

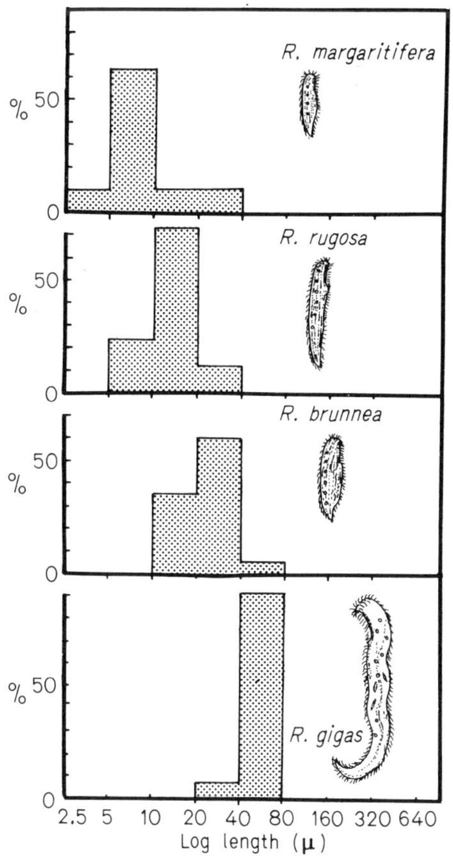


Fig. 40. The size ranges of diatoms eaten by 4 species of ciliates of the genus *Remanella* which frequently coexist in marine sandbottoms. (Redrawn after Fenchel, 1968)

preference are component of the habitat niche. Figure 39 shows the habitat niches of four species of marine isopods.

Species may be active simultaneously in the same habitat but may differ in their utilization of (food) resources; they thus differ with respect to their resource niches. Figure 40 shows the sizes of diatoms eaten by four congeneric species of ciliates coexisting in marine sand. In this case the niche may be described by a resource utilization function which specifies the rate at which a species utilizes the different resource qualities (Fig. 41 and next section).

In actual practice it may be difficult to distinguish between habitat and resource aspects of the niche. Two kinds of fish may pick worms living on and beneath rocks respectively, either because of their morphological or behavioral characteristics making them more suited to collect their food in the particular habitat, or because of their preference for the specific kinds of worms living in the two different habitats. What further complicates the matter is that although two different species have different habitat or time niches, their resources may be correlated. A pair of species, one active at day, the other at night, or the above mentioned habitat selective fishes could in fact exploit the same prey population. Different insects feeding on different parts of the same plant in fact exploit the same resource. This problem has not been analyzed sufficiently from a theoretical point of view.

Species—at least those which are not very related taxonomically—living in the same area usually differ in more than one of the three dimensions of the niche. As seen from Figure 42 a slight separation of niches in each of two dimensions between two species will give a larger overall separation of the niches if the two resource dimensions are uncorrelated. In fact, by a redefinition of the resources ( a turn of the coordinate system in Fig. 42) it is possible to get two independent resource dimensions, one with complete overlap and one with minimal overlap.

An example of a study of the niche packing and coexistence of species in terms of the niche separations within all three dimensions of the niche: resource, habitat and time, is Pianka's (1973) study of lizard communities.

## 3.3 The Formulation of Competition Coefficients

The following theories of limiting similarity of competing species are based on Levin's formulation of the niche applied to one dimension. Here we will consider a resource spectrum of, e.g., food particle sizes (cf. Fig. 40), but similar theories may be developed to study habitat or time dimensions of the niche.

In a continuous resource spectrum, where the resource quality is described by the parameter $\varrho$, the niche of a species is described by the utilization function $\mathscr{U}(\varrho)$ which is the rate by which the individuals use the different resource qualities (Fig. 41). The "rate of use" function $\mathscr{U}$ may be resolved into a "resource picking" function $f$, which is a probability distribution describing the preference of the different resource qualities, and a rate parameter $\omega$ which measures the rate of consumption per individual per time unit:

$$\mathscr{U}(\varrho)=\omega f(\varrho)\,. \tag{3.7}$$

By definition $f$ measures the probability that an individual will consume a given resource quality, so that

$$\int f(\varrho)d\varrho=1\,.$$

With this definition, the total resource used per time unit per individual may be expressed in terms of the resource availability $\mathscr{R}(\varrho)$ as

$$\int \mathscr{U}(\varrho)\mathscr{R}(\varrho)d\varrho=\omega\int f(\varrho)\mathscr{R}(\varrho)d\varrho\,, \tag{3.8}$$

i.e., the resource utilization is proportional to the mean resource availability, with the proportionality factor being the rate of resource picking $\omega$. The carrying capacity for the species must be inversely proportional to the individual resource

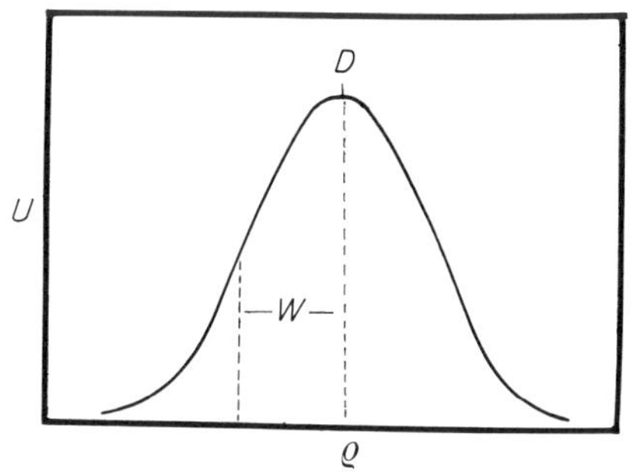


Fig. 41. The utilization function of a species with respect to a gradient in qualities of a resource $\varrho$. This niche is assumed to be normally distributed with the mode $D$ and with the standard deviation or niche width $W$

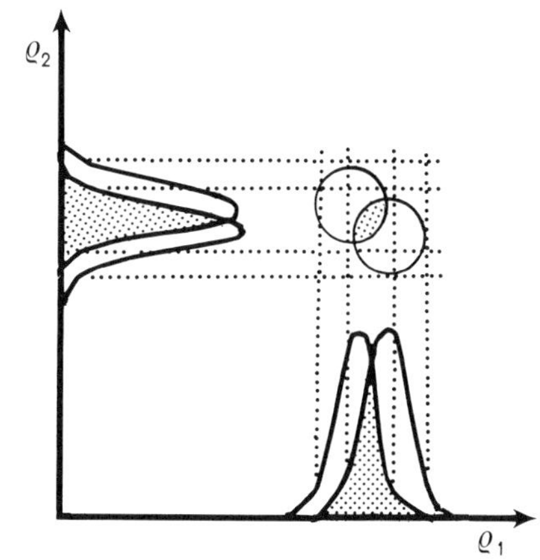


Fig. 42. Two species with a high degree of niche overlap along two niche dimensions. The total niche overlap is less than along either of the two dimensions

use, and therefore proportional to $1/\omega$. Besides this proportionality, the carrying capacity is a function of the mean resource availability.

The definition [Eq. (3.7)] of resource exploitation is admittedly crude. The principal assumption in the formulation is that the species exploit the resources independent of their abundance: The form of $\mathscr{f}$ does not depend on the functional form of $\mathscr{R}$. Similarly, the species exploit the resources with the same intensity whether they are abundant or not, i.e. $\omega$ is independent of the mean resource availability.

If two resource utilization functions of two species, $i$ and $j$ along one niche dimension axis, $\varrho$, partly or totally overlap, then this indicates competition for the resources between the two species (Fig. 43). To measure this competition, consider the expression

$$\beta_{ij} = \int \mathscr{U}_i \mathscr{U}_j d\varrho \Big/ \int \mathscr{U}_i^2 d\varrho \,, \tag{3.9}$$

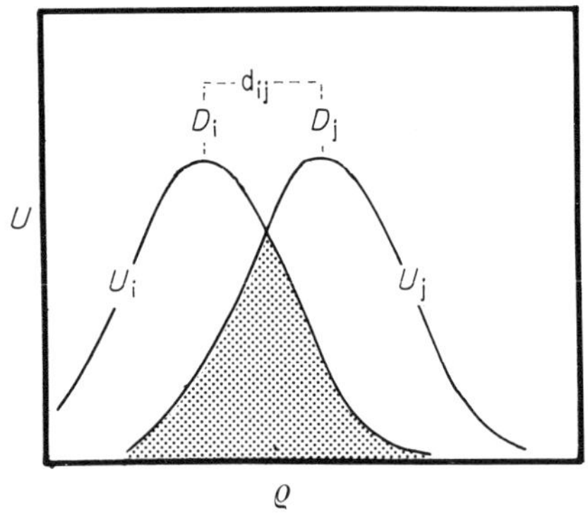


Fig. 43. The partial niche overlap of two species, *i* and *j*, along the resource dimension $\varrho$

which is analogous to Eq. (3.5). This expression is 0 if the curves do not overlap and 1 in case of total identity, i.e., $\beta_{ii}=1$. The expression [Eq. (3.9)] measures the probability of both species utilizing an element of the resource measured by the similar competitive situation for two individuals belonging to species *i*. By this definition of the competition coefficients we may describe the growth of the populations of *m* species utilizing the resources $\varrho$ by Eq. (2.26):

$$dx_i/dt=r_ix_i(K_i-x_i-\sum_{j\neq i}\beta_{ij}x_j)/K_i\,. \tag{3.10}$$

To reach the simple parameterization of Eq. (2.28):

$$dx_i/dt=r_ix_i\left(1-\sum_{j=1}^{m}\alpha_{ij}x_j\right), \tag{3.11}$$

we measure the population size as a fraction of the carrying capacity and $\alpha_{ij}=\beta_{ij}K_j/K_i$. Thus from Eq. (3.9) using the expression [Eq. (3.7)] we have

$$\alpha_{ij}=(K_j/K_i)\,(\omega_i\omega_j/\omega_i^2)\int f_if_jd\varrho/\int f_i^2d\varrho\,. \tag{3.12}$$

From the proportionality of $K_i$ to $1/\omega_i$, we see that the competition coefficient is independent of the basic rate for resource use $\omega_i$. Further, if the mean resource availability is the same for the two species, then the competition coefficient becomes independent of the resource availability:

$$\alpha_{ij}=\int f_if_jd\varrho/\int f_i^2d\varrho\,. \tag{3.13}$$

This situation emerges if all the resource qualities are equally abundant and if the two resource picking functions $f_i$ and $f_j$ are of equal form but of different positions. The consumption rate per individual, $\omega$, only determines the population size in absolute numbers.

Another rationale for the definition [Eq. (3.9)] of the competition coefficients can be achieved by the following considerations. As for the discrete case [Eq. (3.2)] assume that the abundance of the resource is given by

$$\mathscr{R}(\varrho)=\mathscr{S}(\varrho)-x_1\mathscr{U}_1(\varrho)-x_2\mathscr{U}_2(\varrho)\,, \tag{3.14}$$

if we assume that only two species are exploiting the resources. Parallel to Eq. (3.1) the growth of the species is given by

$$dx_i/dt=x_i[\int\mathscr{U}_i(\varrho)\,\mathscr{R}(\varrho)\,d\varrho-\mathscr{T}_i]\,. \tag{3.15}$$

Substituting Eq. (3.14) into Eq. (3.15) yields

$$dx_1/dt=x_1\{[\int\mathscr{S}\mathscr{U}_1d\varrho-\mathscr{T}_1]-x_1\int\mathscr{U}_1^2d\varrho-x_2\int\mathscr{U}_1\mathscr{U}_2d\varrho\}\,, \tag{3.16}$$

which is of the form of Eq. (3.10) with $\beta_{ij}$ given by Eq. (3.9) and

$$\begin{aligned} r_i&=[\int\mathscr{S}\mathscr{U}_id\varrho-\mathscr{T}_i]\,,\\ K_i&=[\int\mathscr{S}\mathscr{U}_id\varrho-\mathscr{T}_i]/\int\mathscr{U}_i^2d\varrho\,, \end{aligned} \tag{3.17}$$

which are analogous to Eq. (3.5)

With the specification [Eq. (3.7)] of the utilization function, the model [Eq. (3.14)] provides from Eq. (3.17) the expression for the carrying capacity of species $i$:

$$K_i=(1/\omega_i)\,[\int\mathscr{S}f_id\varrho-\omega_i\mathscr{T}_i]/\int f_i^2d\varrho\,. \tag{3.18}$$

In the subsequent use of this expression, we will always assume that $\mathscr{T}_i$ is a small quantity, such that we may assume $\mathscr{T}_i=0$ and get a satisfactory feeling of the properties of a more general model. With this assumption, Eq. (3.18) reads

$$K_i=(1/\omega_i)\int\mathscr{S}f_id\varrho/\int f_i^2d\varrho\,, \tag{3.19}$$

and $K_i$ is thus proportional to $1/\omega_i$. Therefore, the competition coefficients $\alpha_{ij}$ from Eq. (3.12) are in this case independent of the *rates* $\omega_i$ with which an individual of the species utilizes the resource, but may be dependent on the amount of available resources which as seen from Eq. (3.19) is a component of $K_i$.

In the following we will consider Eqs. (3.12) and (3.19) as definitions of the competition coefficients and the carrying capacities in terms of the resource preference functions $f_i$ and the resource availability function $\mathscr{S}$. However, by doing so, we have to keep in mind the basic assumptions on how the species utilize the resources, namely that the preference function, $f$, and the rate of exploitation, $\omega$, are independent of the resource availability.

In the general model given by Eqs. (3.9) and (3.10) note that the considerations on stability for competitive systems with symmetric competition matrices in Section 2.5 apply. These arguments apply to any competition matrix of the form

$\beta_{ij}=\Phi_i\Psi_j\gamma_{ij}$ where the matrix $\{\gamma_{ij}\}$ is symmetric (the factor $\Psi_j$ may be absorbed as a scaling factor in the population size and $1/\Phi_i$ is then a factor in the carrying capacity). Thus a valid equilibrium of Eq. (3.10) is stable provided the matrix $\{\int \mathscr{U}_i\mathscr{U}_j d\varrho\}$ or the matrix $\{\int f_i f_j d\varrho\}$ is positive definite.

## 3.4 Limiting Similarity

In the following we will explore the dynamics of a community of species which interact by exploitative competition. We will, throughout the section, assume that the niches are Gaussian, i.e., that $f(\varrho)$ is a normal distribution density. This is often a close approximation to what is found in nature.

The niche described by the utilization function $\mathscr{U}(\varrho)=\omega f(\varrho)$ (where $f$ is a Gaussian distribution function) may be characterized by the mean $D$ of $f$, viz., the *position of the niche*, and by the standard deviation $W$ of $f$, which is called the *niche width*. Now consider two niches $\mathscr{U}_i$ and $\mathscr{U}_j$ in a habitat with a uniform resource spectrum, i.e., $\mathscr{R}(\varrho)$ constant. The competition coefficient Eq. (3.13) may now be expressed in terms of the niche width and the *modal distance* $d_{ij}=D_i-D_j$ (Fig. 43). In the following we will assume that $W_i=W_j=W$, and by this assumption the carrying capacities of the two species are equal save for a factor of the resource consumption rates $\omega_i$ or $\omega_j$. This assumption also simplifies the calculation of the competition coefficients, and by integration of Eq. (3.13) we have that

$$\alpha_{ij}=\exp[-d_{ij}^2/(4W^2)]\,, \tag{3.20}$$

To see this, we first note that the Gaussian distribution has the form:

$$f(\varrho)=[1/\sqrt{(2\pi W^2)}]\exp[-(\varrho-D)^2/(2W^2)]\,. \tag{3.21}$$

Now, the numerator of $\alpha_{ij}$ becomes

$$\begin{aligned}\int f_i f_j d\varrho &= \{1/(2\pi W^2)\}\int \exp[-(\varrho-D_i)^2/(2W^2)]\ \exp[-(\varrho-D_j)^2/(2W^2)]\,d\varrho\\ &= \{1/2\pi W^2)\}\int \exp\{-[(\varrho-D_i)^2+(\varrho-D_j)^2]/(2W^2)\}\,d\varrho\,.\end{aligned} \tag{3.22}$$

The square bracket can be rewritten:

$$(\varrho-D_i)^2+(\varrho-D_j)^2=2[\varrho-(D_i+D_j)/2]^2+(D_i-D_j)^2/2\,. \tag{3.23}$$

which inserted into Eq. (3.22) reveals

$$\int f_i f_j d\varrho=\{1/(2\pi W^2)\}\exp[-(D_i-D_j)^2/(4W^2)]\int \exp\{-[\varrho-(D_i+D_j)/2]^2/W^2\}\,d\varrho\,. \tag{3.24}$$

Here the integral is nearly an integral of a normal distribution like Eq. (3.21) with $D=(D_i+D_j)/2$ and half the variance, the only thing missing is a factor $1/\sqrt{[2\pi(W^2/2)]}$. The integral of a probability distribution is unity, such that the integral on the right side of Eq. (3.24) is $\sqrt{[2\pi(W^2/2)]}$ and we get

$$\int f_i f_j d\varrho=[1/\sqrt{(4\pi W^2)}]\exp[-(D_i-D_j)^2/(4W^2)]\,. \tag{3.25}$$

To get (3.20) just observe that from Eq. (3.25) we have

$$\int f_i^2 d\varrho = 1/\sqrt{(4\pi W^2)}\,, \tag{3.26}$$

and dividing Eq. (3.25) by Eq. (3.26) yields Eq. (3.20).

Now, consider the simple situation of equally spaced niches along the resource axis, so that $d_{ij}=(j-i)\,d$, where $d$ is the modal distance between neighboring species (Fig. 43). In this model [Eq. (3.20)] becomes

$$\alpha_{ij}=\alpha^{(i-j)^2}\,, \tag{3.27}$$

where

$$\alpha=\exp(-d^2/4W^2) \tag{3.28}$$

is the competition coefficient of neighboring species. With this competition matrix, species with distance $2d$ have the competition coefficient $\alpha^4$, species with distance $3d$, $\alpha^9$, etc.

A competition matrix of the form [Eq. (3.27)] is symmetric, so that the results on stability in Section 2.5 apply here. May (1974) showed that a competition matrix of the form [Eq. (3.27)] is always positive definite, so any valid equilibrium in this model will be stable.

Let us first consider a two species system, which has the competition matrix

$$\mathbf{B}=\begin{Bmatrix}1 & \alpha\\ \alpha & 1\end{Bmatrix} \tag{3.29}$$

and, as shown in Chapter 2, p. 59, always gives the stable equilibrium

$$\hat{x}_1=\hat{x}_2=1/(1+\alpha) \tag{3.30}$$

if $\alpha<1$, which is always the case from Eq. (3.28).

For a three species system, where the middle species is subject to diffuse competition (Fig. 44), we have

$$\mathbf{B}=\begin{Bmatrix}1 & \alpha & \alpha^4\\ \alpha & 1 & \alpha\\ \alpha^4 & \alpha & 1\end{Bmatrix} \tag{3.31}$$

which, as shown in Section 2.6, has a stable equilibrium

$$\begin{aligned}\hat{x}_1&=\hat{x}_3=1/[(1-\alpha)\,(1+\alpha)^2]\\ \hat{x}_2&=(1-\alpha-\alpha^2-\alpha^3)/[(1-\alpha)\,(1+\alpha)^2]\end{aligned} \tag{3.32}$$

provided that $1-\alpha-\alpha^2-\alpha^3>0$ or that $\alpha<0.54$, which from Eq. (3.28) yields the condition

$$|d/W|>1.6\,. \tag{3.33}$$

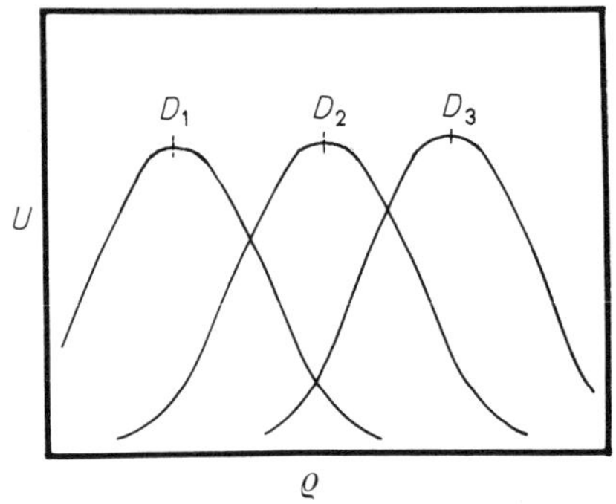


Fig. 44. Diffuse competition with three species

Table 1. Limiting similarity of equally spaced species in an unlimited resource spectrum

| Number of species | 2 | 3 | 4 | 5 | 6 |
|---|---|---|---|---|---|
| $\lvert d/W \rvert >$ | 0 | 1.6 | 1.1 | 1.5 | 1.3 |

For higher numbers of species similar conditions may be found; a summary is given in Table 1. It is seen that in any case, except that of two species, there is a nontrivial lower bound to the closeness of the niches. This lower bound is termed the *limiting similarity* of the species (MacArthur and Levins, 1967).

The considerations above are based on simplifying assumptions, viz., a symmetrical competition matrix (equally spaced niches) and equal carrying capacities and niche widths of all species. Also it has been assumed that the resource spectrum is unlimited. As will be shown in a later section, these assumptions are not always very unrealistic.

If the resources are limited at the borders of the resource spectrum, for example if the resources themselves have a Gaussian distribution, then the conditions for coexistence become somewhat relaxed, because the border species are more resource limited, rarer, and therefore exert less intense competition on the neighboring species. This is in contrast to the situation considered above, where the highest population size is always found in the most extreme species, see Eq. (3.32).

Consider therefore the situation where the resource abundance along the resource axis has a Gaussian shape. In terms of the model discussed on p. 66 this corresponds to $\mathscr{S}$ being a normal distribution function, with variance $\sigma^2$, say, and the mean arbitrarily set to zero. Now the simple form [Eq. (3.13)] of the competition coefficients $\alpha_{ij}$ is no longer valid, rather we have to use the form [Eq. (3.12)] with $K_i$ defined by Eq. (3.17). For convenience assume that the threshold rate for resource consumption, $\mathscr{T}_i$, is low such that we may neglect it by assuming that $\mathscr{T}_i = 0$. i.e.,

$$K_i = \int \mathscr{S} \mathscr{U}_i d\varrho / \int \mathscr{U}_i^2 d\varrho \,. \tag{3.34}$$

By an argument similar to the one leading to Eq. (3.20) we get

$$K_i = (1/\omega_i)(\sqrt{s})\exp\{-D_i^2/[2(\sigma^2+W^2)]\}, \tag{3.35}$$

where

$$s = 2W^2/(\sigma^2+W^2). \tag{3.36}$$

Thus, the competition coefficients are now, from Eq. (3.25), given by

$$\alpha_{ij} = \exp\{(D_i^2 - D_j^2)/[2(\sigma^2+W^2)]\}\exp\{-(D_i-D_j)^2/(4W^2)\}. \tag{3.37}$$

The expression [Eq. (3.37)] for the competition coefficients is inconvenient and not easily interpretable. Define the function

$$k(D) = \exp\{-D^2/[2(\sigma^2+W^2)]\}, \tag{3.38}$$

which from Eq. (3.35) is the carrying capacity $K$ measured relative to the carrying capacity of the species, had its niche position been at the center of the resource [$k(0)=1$]. The form [Eq. (3.11)] of the growth equations for the species was reached by considering the population density as a fraction of the carrying capacity. If we now instead measure the population sizes relative to the carrying capacity the species would have with an optimal niche location ($D=0$), then the growth equations become

$$dx_i/dt = r_i x_i \left[k(D_i) - \sum_{j=1}^{m} \alpha_{ij} x_j\right] \Big/ k(D_i) \tag{3.39}$$

where $\alpha_{ij}$ is given by Eq. (3.20) as in the case of unlimited resources. Thus, this limitation of the resource only has the effect of changing the carrying capacities, not the basic competition structure.

Let us now, for comparison with Eq. (3.33), limit the attention to a three species model where the niches are located at the positions $-d$, 0, and $d$ for species 1, 2, and 3, respectively. This should obviously give the most relaxed limiting similarity condition in the case of limited resources. The competition matrix is again given by Eq. (3.31) and the relative carrying capacities are from Eq. (3.38):

$$k_1 = k_3 = k(d) = \exp\{-d^2/[2(\sigma^2+W^2)]\} \quad \text{and} \quad k_2 = 1, \tag{3.40}$$

which provides the equilibrium solution:

$$\begin{aligned} \hat{x}_1 = \hat{x}_3 &= [k(d)-\alpha]/(1-\alpha^2)^2, \\ \hat{x}_2 &= [1+\alpha^4 - 2\alpha k(d)]/(1-\alpha^2)^2. \end{aligned} \tag{3.41}$$

Table 2. Limiting similarity for three species dependent on a Gaussian resource

| $\sigma/W$ | 1.1 | 1.4 | 1.7 | 2 | 3 | 4 | 5 | $\infty$ |
|---|---|---|---|---|---|---|---|---|
| $\lvert d/W \rvert >$ | 0.4 | 0.8 | 1.0 | 1.1 | 1.3 | 1.4 | 1.5 | 1.6 |

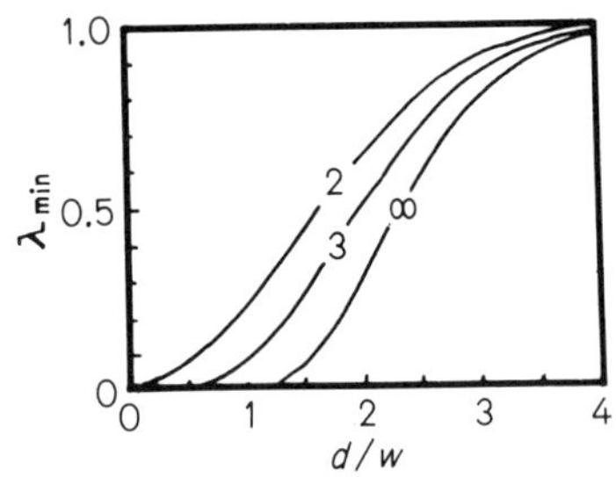


Fig. 45. The minimum eigenvalue of the competition matrix as function of $d/W$ for systems with 2, 3, and infinitely many competing species. (Redrawn from May, 1973a)

(MacArthur and Levins, 1967). The conditions for validity of this equilibrium are first that $k(d) > \alpha$, which is equivalent to the condition

$$\sigma > W \tag{3.42}$$

i.e., that the resource "width" should be greater than the niche width, while the second condition is the parallel of Eq. (3.33):

$$1 + \alpha^4 > 2\alpha k(d) . \tag{3.43}$$

This condition is weaker than Eq. (3.33), which corresponds to $k(d) = 1$, and here $k(d) < 1$. For $\sigma \to W$ (i.e., $\hat{x}_1 \to 0$) it becomes true for all $\alpha$, but as seen from Table 2 the condition rapidly becomes severe for increasing $\sigma$ and the limiting similarity condition is of the same order of magnitude as shown in Table 1. Thus the conditions in the uniform resource case are not severely biased by the edge effect, i.e., that the border species in the community has a comparatively high population size. A further discussion of this point with general values of the carrying capacities and with more species is given by May (1973a).

The significance of these considerations is that they predict what we in fact find in nature, namely that the limiting similarity, i.e., the minimal value of $d/W$, should be greater than about one for coexisting species. In other words, the niches of coexisting species competing for a limiting resource are displaced along the resource axis by approximately one niche width unit (Figs. 40, 51, 53).

Another line of reasoning giving the same result is due to May (1973a, 1974). It is based on the observation that the absolute value of the minimum eigenvalue (rate of return) around the equilibrium essentially vanishes for $d/W$ in an interval above zero (Fig. 45). Therefore, for large values of $n$ (already $n \geqq 3$) the return to equilibrium is extremely slow for $0 \leq d/W \leq 1$ so that environmental stochasticity may become significant (see Chap. 4). The environmental variation may be described by a random noise around the mean carrying capacity $\bar{K}$ with a variance $s^2$. The requirement for stability is then roughly that the minimum eigenvalue of the competition matrix should exceed $s^2/\bar{K}$ (May, 1973a). What is important here, is that the minimum eigenvalue should be bounded away from zero. But the eigenvalue becomes practically zero for $d/W$ less than unity, so we get the requirement for stability that $d/W$ should be greater than about one. May's (1973a) results are questioned by Feldman and Roughgarden (1975) (for further discussion see Chap. 4).

The significance of May's (1973a, 1974) result is restricted by the fact that except for a two species competitive system the deterministic models discussed above show that the niches cannot be packed closer than $d/W = 1$ anyway. In all circumstances, however, we conclude that considerations based on admittedly simplified assumptions lead to the celebrated limiting similarity condition that $d/W$ should be greater than about unity.

May (1974) and Roughgarden (1974b) consider other shapes of the utilization function. May (1974b) gives examples of both thick and thin tailed utilization functions which produce positive definite competition matrices, all with the property that the minimum eigenvalue vanishes in an interval of $d/W$ above zero. Roughgarden (1974b) uses the condition for invasion of the middle species in a three species community to determine the limiting similarity. Under symmetric competition, this invasibility condition is, as seen in Chapter 2, p. 57, the condition for existence of a stable three species equilibrium. Roughgarden (1974b) concludes that a community where the species have thick tailed utilization functions can be tighter packed than a community with thin tailed utilization functions. This is concluded by measuring the limiting similarity as a bound on $d/W$ where $W$ is the square root of the variance of the utilization function. However, this width parameter may not exactly correspond to what is intuitively thought of as amount of specialization of the species. Two species, each having a niche width of $W$, but one with a thick and one with a thin tailed use of the resource, have a quite different "width" of the resource qualities that they specialize on. Thus, viewed superficially, the packing in the two communities may be quite similar.

## 3.5 The Evolution of Competing Species

It is easy to make intuitive and verbally formulated models indicating that in a community of competing species, natural selection will tend to relax competition by separating the niches, i.e., by separating the utilization functions. Where a species lives together with a competitor, individuals of a phenotype which utilizes the part of the resource spectrum not used by the competitor will have a higher fitness than an individual of a phenotype which uses the same part of the resources as the competing species. It is, perhaps, also intuitively acceptable that selection will tend to equalize modal distances and niche widths of species involved in diffuse competition.

Table 3. Internal structure of the population of species $y$

| Genotype | $AA$ | $Aa$ | $aa$ | $\sum$ |
|---|---|---|---|---|
| Number | $y_1$ | $y_2$ | $y_3$ | $y$ |
| Frequency | $p^2$ | $2pq$ | $q^2$ | 1 |
| Competition coefficients | $\alpha_{x1}$ | $\alpha_{x2}$ | $\alpha_{x3}$ | — |
| | $\alpha_{z1}$ | $\alpha_{z2}$ | $\alpha_{z3}$ | — |

It is much more difficult to make mathematially formulated models which show how selection will act on populations with Mendelian genetics involved in competition. Below we suggest such models which under simplified assumptions may be analyzed and show that selection will generally tend to minimize exploitative competition (see also Roughgarden, 1975). In the following section we will discuss some of the assumptions in more detail in connection with a discussion on the components of the niche.

Our basic set up is a community of three species, $x$, $y$, and $z$, arranged in that order along a resource axis. We will assume discrete generations, and that the species grow according to the following difference equations in a uniform resource spectrum:

$$\begin{aligned} x'-x &= Rx(1-x-\alpha_x y-\beta z)\,, \\ y'-y &= Ry(1-\alpha_x x-y-\alpha_z z)\,, \\ z'-z &= Rz(1-\beta x-\alpha_z y-z)\,, \end{aligned} \tag{3.44}$$

where $R$ is the finite growth rate and the left sides of the equations are the increments of the populations in one generation (see Sect. 1.1).

Assume that in species $y$ there is genetic variation influencing the competition with $x$ and $z$, and let us for simplicity assume that this variability is at a single autosomal locus with two alleles $A$ and $a$, which have the frequencies $p$ and $q$ $(=1-p)$. The structure of the model is summarized in Table 3 and in terms of these parameters, the parameters in Eq. (3.44) are given by

$$\begin{aligned} \alpha_x &= p^2\alpha_{x1}+2pq\alpha_{x2}+q^2\alpha_{x3}\,, \\ \alpha_z &= p^2\alpha_{z1}+2pq\alpha_{z2}+q^2\alpha_{z3}\,. \end{aligned} \tag{3.45}$$

The numbers of each of the genotypes of species $y$ change through one generation to

$$y_i'-y_i=Ry_i(1-\alpha_{xi}x-y-\alpha_{zi}z) \tag{3.46}$$

so that the change in gene frequency is

$$p'-p=[(y_1'+y_2'/2)/y']-p=[(y_1'+y_2'/2)-py']/y'\,. \tag{3.47}$$

Consider first the situation of two competing species with $z=0$. Substituting the expressions for $y_i'$ and $y'$ from Eq. (3.46) and Eq. (3.44) into Eq. (3.47) yields the equation

$$p'-p=-R(y_1x\alpha_{x1}+y_2x\alpha_{x2}/2-pyx\alpha_x)/y' \tag{3.48}$$

which, by substituting $\alpha_x$ from Eq. (3.45) and the genotypic frequencies from Table 3, produces

$$p'-p=(-Rpqxy/y')\,[p(\alpha_{x1}-\alpha_{x2})+q(\alpha_{x2}-\alpha_{x3})]\,. \tag{3.49}$$

By arguments similar to the ones arriving at Eq. (1.34), (in this case by differentiating Eq. (3.45) with respect to $p$) we can now from the form of Eq. (3.49) deduce that selection will tend to decrease $\alpha_x$; in fact,

$$p'-p=[-Rpqxy/(2y')]\,d\alpha_x/dp\,. \tag{3.50}$$

From the expression Eq. (3.20) for the competition coefficient, we see that the only ways in which $\alpha_x$ can decrease is by increasing $d_{xy}$ or by decreasing $W$, that is by increasing niche separation $d_{xy}/W$. In other words the niches will diverge so we have given an example of "character displacement".

To find the equilibria of the two species system with genetic variation, we set Eq. (3.44) and Eq. (3.49) equal to zero and find that

$$\hat{x}=\hat{y}=1/(1+\alpha_x) \tag{3.51}$$

and $\hat{p}=0$ or $\hat{p}=1$ or

$$\hat{p}=(\alpha_{x2}-\alpha_{x3})/(2\alpha_{x2}-\alpha_{x1}-\alpha_{x3})\,, \tag{3.52}$$

with $\alpha_x$ in each case given by Eq. (3.45). The equilibrium in the population sizes is stable when $\alpha_x<1$ and if each of the two species have a stable equilibrium when they live alone (cf. Sect. 2.2). The genetic equilibrium given by Eq. (3.52) is valid and stable if the competition coefficient of the heterozygote is lover than that of either homozygote, corresponding to overdominance in ordinary fitness parameters (Sect. 1.2). Similarly Eq. (3.52) is valid, but unstable in the case of underdominance. Otherwise, the equilibrium Eq. (3.52) does not exist and the population will be fixed on the allele with the lowest competition coefficient.

Note that in this model we have ignored changes in the intraspecific competition; we will return to that question later and only note that if the difference in niche position of the two alleles is small compared with the distance between the species and if the niche separation between the two species is not too large, then the above analysis provides a valid approximation. Similarly, we have not considered the possibility that the resource availability is limited in the peripheries of the resource spectrum of the two species.

Now, let us return to the general three species model [Eq. (3.44)] with the genetic variation in species $y$ as shown in Table 3. Two questions may be asked in this model: (1) For a fixed niche width $W$, how will $d_{xy}$ and $d_{yz}$ change, and (2) for a fixed $d_{xy}=d_{yz}=d$, how will $W$ change.

By a method similar to that used for arriving at Eq. (3.49) we find that the gene frequency change in the three species model is

$$p' - p = (-Rpqy/y')\,[p(\alpha_1^* - \alpha_2^*) + q(\alpha_2^* - \alpha_3^*)] \tag{3.53}$$

where

$$\alpha_i^* = x\alpha_{ix} + z\alpha_{iz}\,. \tag{3.54}$$

Equation (3.53) for the gene frequency change looks like Eq. (3.49), but is much more complicated because of Eq. (3.54), which shows that the relation between the "fitness" parameters is determined by the population sizes. Thus we cannot use the simple procedure from the two species system and no attempt to find a general solution will be made. However, we can in any case determine the conditions for initial increase of a rare allele in the population.

Suppose that the population consists primarily of individuals of genotype *aa* with a low frequency of the allele *A* in the population. If in this situation we neglect terms of the order $p^2$ in Eq. (3.53) then we get approximately

$$p' - p = (-Ry/y')\,p(\alpha_2^* - \alpha_3^*) \tag{3.55}$$

so that the rare allele *A* will increase in the population provided

$$\alpha_2^* < \alpha_3^*\,, \tag{3.56}$$

i.e., *A* increases when the competition experienced by the heterozygote is less severe than that of the predominant homozygote.

To evaluate Eq. (3.56) we may assume that the population sizes are at the equilibrium

$$\begin{aligned} \hat{x} &= (1-\alpha_z)\,[(1-\beta) - (\alpha_x - \alpha_z)]/D\,, \\ \hat{y} &= (1-\beta)\,[1+\beta - (\alpha_x + \alpha_z)]/D\,, \\ \hat{z} &= (1-\alpha_x)\,[(1-\beta) + (\alpha_x - \alpha_z)]/D\,, \end{aligned} \tag{3.57}$$

where $D = (1-\beta)(1+\beta-2\alpha_x\alpha_z) - (\alpha_x - \alpha_z)^2$, and the equlibrium is valid if

$$\alpha_x + \alpha_z < 1 + \beta \tag{3.58}$$

and $\beta < \alpha_x, \alpha_z < 1$, and it is always stable for *R* small.

Consider now the example in Figure 46. Each species has a niche width *W*, but the modal distance between *x* and $y(=d+e)$ differs from that between *y* and $z(d-e)$. By Eq. (3.20) we get the competition coefficients in a population with only *aa* as

$$\begin{aligned} \alpha_{x3} &= \exp[-(d+e_3)^2/(4W^2)]\,, \\ \alpha_{z3} &= \exp[-(d-e_3)^2/(4W^2)]\,, \\ \beta &= \exp[-(2d)^2/(4W^2)]\,, \end{aligned} \tag{3.59}$$

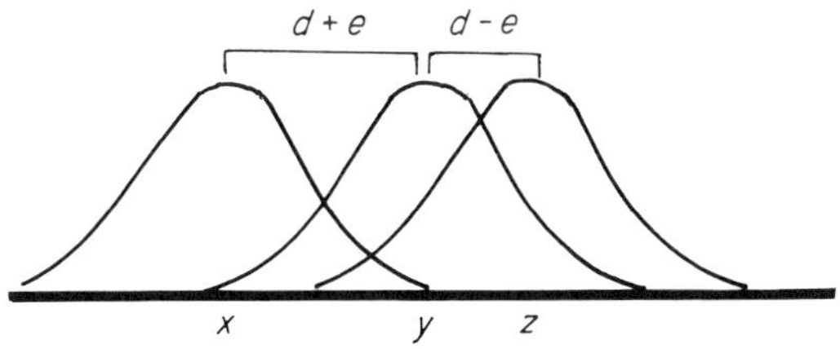


Fig. 46. Diffuse competition with three species; for fixed $D_x$ and $D_z$, $D_y$ will move towards the middle ($e \to 0$) as the result of natural selection

and the competition coefficients of the rare heterozygote as

$$\alpha_{x2} = \exp[-(d+e_2)^2/(4W^2)] = \exp\{-[(d+e_3)+(e_2-e_3)]^2/(4W^2)\},$$
$$\alpha_{z2} = \exp[-(d-e_2)^2/(4W^2)] = \exp\{-[(d-e_3)-(e_2-e_3)]^2/(4W^2)\}. \quad (3.60)$$

If we assume that the niche position of the two genotypes, *aa* and *Aa*, only differs slightly, then we can get a good approximation of Eq. (3.60) by Taylor expansion around $e_2 - e_3$ and neglecting square terms. By this procedure we get the approximation

$$\alpha_{x2} = \alpha_{x3}[1-(e_2-e_3)(d+e_3)/(2W^2)],$$
$$\alpha_{z2} = \alpha_{z3}[1+(e_2-e_3)(d+e_3)/(2W^2)]. \quad (3.61)$$

Using this, we can write the condition [Eq. (3.56)] for increase of $A$ as

$$0 < (e_2-e_3)[x\alpha_{x3}(d+e_3) - z\alpha_{z3}(d-e_3)]$$

which simplifies to

$$0 < (e_2-e_3)[(x\alpha_{x3} - z\alpha_{z3})d + (x\alpha_{x3} + z\alpha_{z3})e_3]. \quad (3.62)$$

Thus the condition for increase of $A$ in terms of the sign of $(e_2 - e_3)$ is determined by the sign of the bracket. From the equilibrium population sizes given by Eq. (3.57) by setting $\alpha_x = \alpha_{x3}$ and $\alpha_z = \alpha_{z3}$, we get after some calculations

$$D(\hat{x}\alpha_{x3} - \hat{z}\alpha_{z3}) = (\alpha_{x3} - \alpha_{z3})[(1-\beta) - (\alpha_{x3} + \alpha_{z3}) + 2\alpha_{x3}\alpha_{z3}] \quad (3.63)$$

(where the second factor is positive and the first factor is of sign $-e_3$), and

$$D(\hat{x}\alpha_{x3} + \hat{z}\alpha_{z3}) = (\alpha_{x3} + \alpha_{z3} - 2\alpha_{x3}\alpha_{z3})(1-\beta) - (\alpha_{x3} - \alpha_{z3})^2. \quad (3.64)$$

Now, these expressions are still too complicated to handle, but we may analyze the situation in the case where $e_3$ is initially small compared to $d$. In this situation condition [Eq. (3.62)] may be rewritten into Eq. (3.67).

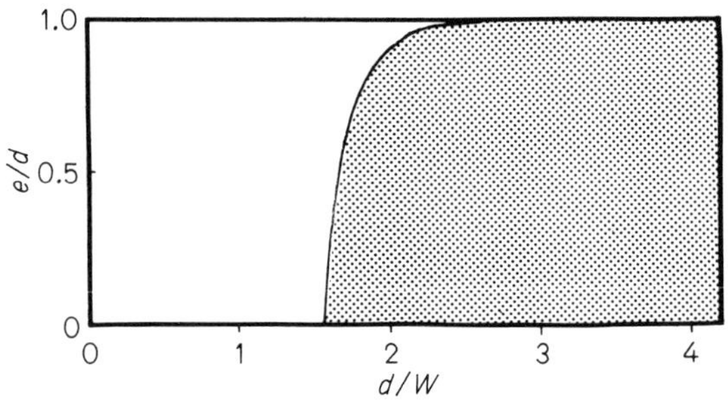


Fig. 47. The region (the shaded area) with a stable and valid equilibrium of a system as shown on Figure 46 as a function of $e/d$ and of $d/W$

By Taylor expansion of Eq. (3.59) we get the approximation:

$$\alpha_{x3} = \alpha[1 - e_3 d/(2W^2)],$$
$$\alpha_{z3} = \alpha[1 + e_3 d/(2W^2)], \tag{3.65}$$

where $\alpha$ is given by Eq. (3.28). From this we get as an approximation

$$\alpha_{x3} - \alpha_{z3} = -e_3 \alpha d/(W^2), \tag{3.66}$$

showing that both terms in the bracket in Eq. (3.62) are approximately proportional to $e_3$. This simplifies the evaluation of that bracket, because its sign is then determined by pulling out a factor $e_3$ and setting $e_3 = 0$ in the rest of the terms. Thus, using Eqs. (3.63) and (3.64) we get Eq. (3.62) to be

$$0 < (e_2 - e_3) e_3 [-\alpha(1 - \beta' - 2\alpha + 2\alpha^2)(d/W)^2 + 2(1 - \beta)\alpha(1 - \alpha)]$$

which by using $\beta = \alpha^4$ simplifies to

$$0 < (e_2 - e_3) e_3 \alpha(1 - \alpha)(1 - \alpha + \alpha^2 + \alpha^3)[2(1 - \alpha^4)/(1 - \alpha + \alpha^2 + \alpha^3) - (d/W)^2]. \tag{3.67}$$

Calculations of the values of the functions in the bracket show that for all $\alpha$ less than 0.54 (the limiting similarly condition, see Table 1) the bracket is negative. Thus the condition for increase of the rare allele $A$ is $e_2 < e_3$, i.e., the niche of the genotype $Aa$ should be closer to the midpoint between the niches of the species $x$ and $z$. In fact, direct numerical calculations on Eq. (3.62) show that this conclusion is not restricted to the situation where $e_3$ is small, but extends to any value of $e_3$ for wich the equilibrium [Eq. (3.57)] is valid (Fig. 47). The convergence of the niche of the middle species towards the midpoint position is not nesessarily just a small adjustment; as Figure 47 shows the change may be quite considerable.

Let us now consider the question of the evolution of the variance of species $y$. The variance of the utilization function of a species may change in two qualitatively different ways, either by increasing the variance in the resources utilized by an individual or by incorporating variation in resource use between the individuals (Roughgarden, 1972, 1974a). To separate the selection for change in variance from the effects previously discussed in this section we will here only

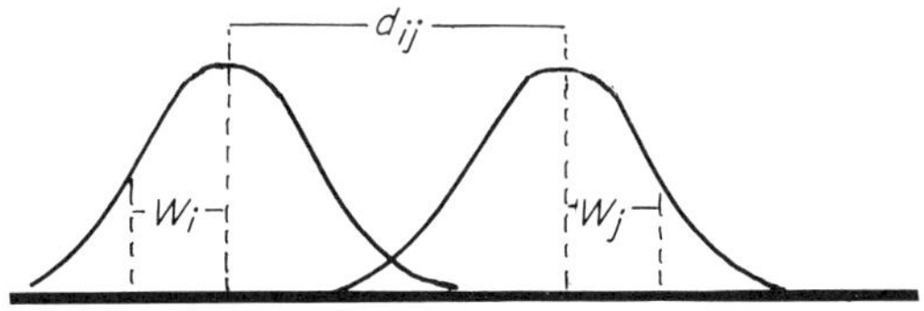


Fig. 48. A system consisting of two competing species; given an infinite resource spectrum, natural selection will tend to increase $d_{ij}$

consider the three species situation where the niches are placed with a distance $d$ between neighbors.

To explore the evolution of the variance in resource utilization between individuals in a population, let us consider the situation where the modes of the utilization functions of the three genotypes are equally spaced with a distance $\varepsilon$. Polymorphism at the locus will then increase the utilization variance of species $y$. However, we already know, that interspecific competition will favor the genotype whose niche position is closest to the midpoint between the niches of the species $x$ and $z$, so an increase in variance will result if that genotype is the heterozygote. On the other hand, if the optimal genotype is a homozygote the intraspecific competition will be relaxed by the variation within the species, such that polymorphism may still be possible. The competition coefficient $\gamma_{ij}$ between genotype $i$ and genotype $j$ is given by the matrix

$$\mathbf{\Gamma}=\left\{\begin{matrix} 1 & \gamma & \gamma^4 \\ \gamma & 1 & \gamma \\ \gamma^4 & \gamma & 1 \end{matrix}\right\} \tag{3.68}$$

where

$$\gamma=\exp(-\varepsilon^2/4W^2)\,, \tag{3.69}$$

as for symmetric competition between species. If we assume that $\varepsilon$ is small as before, then we get the approximation

$$\gamma=1-\varepsilon^2/(4W^2)\,. \tag{3.70}$$

Thus the displacement of the niche of *Aa* by the amount $\varepsilon$ relative to the niche of *aa* only relaxes the interspecific competition by an amount proportional to $\varepsilon^2$. This is in contrast to the change in interspecific competition which from Eq. (3.61) is of the order of $\varepsilon=e_2-e_3$. The previous assumption, that the change in intraspecific competition is negligible in the consideration of the adjustment of the niche position under diffuse competition is therefore justified. However, when the center species, $y$, is initially at the midpoint between $x$ and $z$, then $x=z$ and the interspecific competition pressure experienced by the new genotype *Aa* is proportional to $\alpha_{x2}+\alpha_{z2}$, which from Eq. (3.61) is independent of the first order term in $\varepsilon$. Therefore, the difference between *aa* and *Aa* in inter- and intraspecific competition becomes of the same order of magnitude and we have to allow for the intraspecific competition coefficients to become less than unity.

The intraspecific competition in species $y$ does not alter the growth equations for species $x$ and $z$ in Eq. (3.44), but the growth of the genotypes in species $y$ in Eq. (3.46) has to be altered such that

$$y_i' - y_i = R y_i \left(1 - \alpha_{xi} x - \sum_{j=1}^{3} \gamma_{ij} y_j - \alpha_{zi} z\right), \tag{3.71}$$

and the growth of the total population size of species $y$ is given by

$$y' - y = R y \left(1 - \alpha_x x - \sum_{i=1}^{3} \sum_{j=1}^{3} \gamma_{ij} y_i y_j / y - \alpha_z z\right). \tag{3.72}$$

The change in gene frequency, given by Eq. (3.53) is now a bit more complicated. It turns out to be

$$p' - p = (-Rpqy/y')\{[p(\alpha_1^* - \alpha_2^*) + q(\alpha_2^* - \alpha_3^*)] + [p(\bar{\gamma}_1 - \bar{\gamma}_2) + q(\bar{\gamma}_2 - \bar{\gamma}_3)]y\} \tag{3.73}$$

where

$$\bar{\gamma}_i = p^2 \gamma_{i1} + 2pq\gamma_{i2} + q^2 \gamma_{i3} \tag{3.74}$$

and $\alpha_i^*$ is given by Eq. (3.54). Assume as before that the allele $A$ is rare, so that we can neglect terms of the order $p^2$ to get the condition for increase of $A$. From Eqs. (3.73) and (3.74), the condition for increase of $A$ becomes

$$(\alpha_2^* - \alpha_3^*) + y(\gamma_{23} - \gamma_{33}) < 0. \tag{3.75}$$

Now assume that the niche position of the genotype $aa$ is exactly at the midpoint between $x$ and $z$. This implies that $\alpha_{x3} = \alpha_{z3} = \alpha$ and $\beta = \alpha^4$, such that in a population where $aa$ is the only genotype we have $\hat{x} = \hat{z}$. The competition coefficients of the new genotype $Aa$ become

$$\begin{aligned} \alpha_{x2} &= \exp[-(d+\varepsilon)^2/(4W^2)], \\ \alpha_{z2} &= \exp[-(d-\varepsilon)^2/(4W^2)], \end{aligned} \tag{3.76}$$

which by neglecting the cubic term in the Taylor expansion is approximated by

$$\begin{aligned} \alpha_{x2} &= \alpha[1 - \varepsilon d/(2W^2) + \varepsilon^2 d^2/(8W^4) - \varepsilon^2/(4W^2)], \\ \alpha_{z2} &= \alpha[1 + \varepsilon d/(2W^2) + \varepsilon^2 d^2/(8W^4) - \varepsilon^2/(4W^2)], \end{aligned} \tag{3.77}$$

so that

$$\alpha_{x2} + \alpha_{z2} = \alpha[2 + \varepsilon^2 d^2/(4W^4) - \varepsilon^2/(2W^2)]. \tag{3.78}$$

With this approximation and Eq. (3.70), the condition [Eq. (3.75)] is

$$[\varepsilon^2/(4W^2)][\alpha\hat{x}(d/W)^2 - 2\alpha\hat{x} - \hat{y}] < 0. \tag{3.79}$$

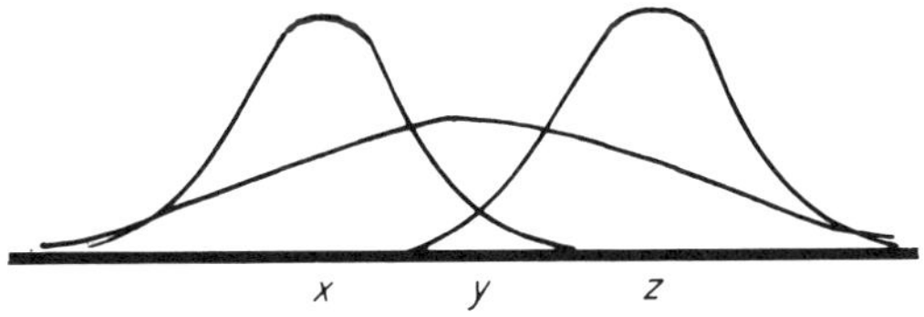


Fig. 49. Diffuse competition with three species; for fixed values of the $D$'s and of $W_x$ and $W_z$, natural selection will tend to decrease $W_y$

But at equilibrium where, e.g., $y' = y$, we have from Eq. (3.44) that $2\alpha\hat{x} + \hat{y} = 1$, so that condition [Eq. (3.79)] is

$$(d/W)^2 < 1/(\alpha\hat{x}) \quad \{= (1 - 2\alpha + \alpha^4)/[\alpha(1-\alpha)]\}. \tag{3.80}$$

This condition is fulfilled for $\alpha < 0.08$ or $d/W > 3.2$, i.e. for loose packing of the niches. Thus when the distance between the niches is sufficiently great, the middle species will increase its utilization variance by incorporating new alleles.

Finally, consider the model illustrated by Figure 49, where the three species are equally spaced, i.e. $d_{xy} = d_{yz} = d$ and $W_x = W_z = W$, but where the individual variance in resource use in species $y$, $W_y$, is subject to genetic variation, such that the three genotypes have the individual variances $W_1$, $W_2$, and $W_3$, respectively. By a calculation similar to the one leading to Eq. (3.35) we may find the competition coefficients of this model by

$$\int f_x f_{yi} d\varrho \Big/ \int f_{yi}^2 d\varrho = [2W_i^2/(W_i^2 + W^2)]^{(1/2)} \exp[-d^2/2(W_i^2 + W^2)] \tag{3.81}$$

which by observing that the carrying capacity as a function of the variance obeys the relation

$$K(W_i)/K(W) = (W_i/W)\,(\omega/\omega_i) \tag{3.82}$$

[from Eq. (3.17) with $\mathscr{S} \equiv 1$ and $\mathscr{T}_i = 0$], so that Eq. (3.12) provides the competition coefficients as

$$\alpha_{xi} = (\sqrt{s_i}) \exp\{-d^2/[2(W_i^2 + W^2)]\} = (\sqrt{s_i})\alpha^{s_i}, \tag{3.83}$$

where

$$s_i = 2W^2/(W_i^2 + W^2). \tag{3.84}$$

Similarly, we get the intraspecific competition coefficients as

$$\gamma_{ij} = [2W_j^2/(W_i^2 + W_j^2)]^{(1/2)}, \tag{3.85}$$

since distance between the modes of the genotypes is zero. Let us restrict attention to the case where the population of $y$ is initially only consisting of genotype $aa$ with $W_3 = W$, and investigate the condition [Eq. (3.75)] for the increase of the allele $A$ which supposedly changes the variance. Then $\gamma_{33}=1$ and $\gamma_{23}=\sqrt{s_2}$ so that Eq. (3.75) becomes

$$(\hat{x}+\hat{z})(\sqrt{s_2}\alpha^{s_2}-\alpha)+\hat{y}(\sqrt{s_2}-1)<0\,, \tag{3.86}$$

which for $s_2$ close to one ($\gamma_{23}\simeq\gamma_{33}$) is approximately

$$\{2\alpha\hat{x}[(1/2)+\log(\alpha)]+\hat{y}/2\}(1-s_2)>0\,. \tag{3.87}$$

As before we have $2\alpha\hat{x}+\hat{y}=1$, so that Eq.(3.87) simplifies to

$$[1+4\alpha\hat{x}\log(\alpha)](1-s_2)>0\,. \tag{3.88}$$

With $\hat{x}$ given by Eq.(3.32), the bracket is a function of $\alpha$ and it turns out to be positive if $\alpha<0.15$ or $d/W>2.8$ and negative elsewhere when the initial equilibrium exists. Therefore, as in our previous model, selection for increased variance ($s_2<1$) only occurs for large distance between the niches. In any other case, selection will favor a decrease in the individual variance.

In the three species models we have only considered the evolution of the middle species with respect to niche position and niche width. However, the situation for the outer species is pretty much that of the two species model considered in the beginning of this section. We would then predict that if genetic variation for the niche parameters occurs in the outer species, then they will show character divergence, and then after some time of divergence, niche expansion too. These phenomena are unbounded in their evolution in the models considered so far, because of the assumption of unlimited resource spectrum, whereas it is evident that divergence of the niches must have a bound if we, for example, have a Gaussian resource spectrum.

The theory for three species evolving in a Gaussian resource spectrum is found in Roughgarden (1975), where he considers the invasion sequence depicted in Figure 31. The first species inhabiting the environment can always sustain a population, but through time it will tend to move to the optimum niche location, where the mode of the resource spectrum and the mode of the utilization function are equal. This evolutionary process is parallel to the convergence of the center species in Figure 47 towards the midpoint between its competitors, and selection depends linearly on the distance of the niche mode from the mode of the resource spectrum. After reaching this goal, the species will start expanding its niche towards the optimum niche width equal to the resource width. This increase in variance occurs by much weaker selection forces, which are of the order of the square of the change of the niche location, parallel to the situation in the middle species in a community of three species. However, the increase in the genetic component of the niche width is possible with the same genetic variation that made the convergence towards the mode of the resource curve possible. Given that the first species in our community has reached its optimum position, then a second species can invade as long as the niche width of the resident species is less than the resource width, i.e., $W+G<\sigma$, where $G$ is the genetic component of the niche width. At this point in time the phenomenon of character divergence may be observed. Here two situations may be distinguished. In the first one, the resident species has had sufficient time for the incorporation of genetic variance, i.e. $G>0$; then if we assume that the immigrant species is more homogeneous, the resident species will show a rather rapid shift in its niche position and a decrease in niche width due to selection against the genotypes whose utilization mode is closer to the mode of the immigrant species. Thus, following this early stage in the two species community, the former resident species may well occupy a more extreme niche position than the immigrant species. The second possibility is that the new species migrates into the habitat before the resident species has expanded its niche significantly. This situation is parallel to the situation after the initial character divergence in the first considered case. Now, given enough variation in the two species, they will tend to get symmetric positions around the resource mode and the distance from the mode will eventually attain an optimum. At this point in time, we may introduce the third species. It will, as we have seen in the previous section, only be able to invade under certain conditions of similarity to the two resident species if we assume that its niche position is between that of the resident species. If the invader succeeds, then processes like the ones considered above come into action. The outer species will show character divergence (rapid if they have built up a genetic component of the niche width) and the middle species will seek the midpoint between its competitors (which in this model also may be the mode of the resource spectrum).

The bricks to build this theory are given by the formulae [Eqs. (3.38) and (3.39), which by using the theory of this section, provide the conditions for increase of a rare allele, e.g., corresponding to Eq. (3.75) we get

$$[(\alpha_2^*)/k(D_2)-(\alpha_3^*)/k(D_3)]+y[\gamma_{23}/k(D_2)-\gamma_{33}/k(D_3)]<0\,. \tag{3.89}$$

Roughgarden (1975) used another line of reasoning corresponding to the reasoning leading to Eq. (1.73). He established the existence of an optimum niche location and an optimum niche width, but the procedure is not suited for the investigation of possibilities of genetic polymorphism.

The arguments presented so far pertain to genetic variation in the exploitation of the resources. With variation of an antagonistic nature the results are quite different. Suppose we have two species with genetic variation of the form of Table 3 in species $y$, but now it is not in the niche position, but in other inhibitive functions of the individuals belonging to species $y$ on species $x$, such that the growth of species $y$ is governed by the equations

$$\begin{aligned} x'-x&=Rx(1-x-\alpha_x y) \quad \text{and} \\ y_i'-y_i&=Ry_i(1-\alpha x-y)\,. \end{aligned} \tag{3.90}$$

Here there will be no change in gene frequency in species $y$, despite the fact that the equilibrium population size of $y$ will be maximal for $\alpha_x$ maximal, because the genotypes all grow according to identical equations (Roughgarden, 1975). However, antagonistic mechanisms may be selected for in geographically structured populations in migration-extinction equilibrium (Chap. 4, p. 120) where the above considered variation in competition functions may be important in determining the successful colonization of a patch and may influence the fitness of the individuals through altering the possibility that the successors will grow and colonize new patches.

## 3.6 Components of the Niche

In order to interpret observations from nature in the light of the result of this chapter, it is necessary to understand the components of the niche in terms of populations and individuals.

In a community of competing species the utilization functions are properties of populations. As pointed out by Fenchel (1975b) and by Roughgarden (1972, 1974a) the niche width is composed of (1) a variance of each individual in the utilization of the resources or in the response to environmental factors and (2) a variance due to variation in the mean between different individuals within the population. These two components have been termed by Roughgarden (1974a) the *within phenotype component*, $V_w$, and the *between phenotype component*, $V_b$. To these components should be added the variance due to age structure of the population, which we may call the *age structure component*, $V_a$. We can therefore formally write

$$\text{Total Niche Width} = V_t = V_w + V_b + V_a\,. \tag{3.91}$$

Each of these components may again have an environmental and a genetical component which is subject to natural selection in different ways and with different rates.

The component of the niche width which is due to the presence of different age groups in a population, $V_a$, is in many ways the most complex one. In many organisms, for example higher vertebrates, this component will only contribute to a somewhat increased variance in, e.g., prey size relative to what would be found if only adult individuals are studied. However, in organisms where metamorphosis is involved in development, e.g., a benthic clam with planktonic larvae, analysis of the true niche width will be extremely complicated. Note that selection on the $V_a$ component will mainly involve changes in the time niche of an organism, e.g., by changing the breeding time or the rate of development (Fenchel, 1975b).

In extreme, a population could have a niche, the variance of which is 100% $V_w + V_a$. For example, a clonal culture of a protozoan in a homogeneous environment would have a niche width due to the variance of individual utilization of the resources plus the variance due to the age structure. The limits to the latter would be set by the difference between a cell ready to divide and its daughter cells which will be of half the size.

The between phenotype component, $V_b$, of the niche is commonly reflected in measurable characters of relevance to the exploitation of the resource, e.g. the size of mouth or beak. This kind of variation is well studied and normally shows an environmental and a genetical component. The question about what regulates the latter is essentially the question of how and why genetical variation is maintained in natural populations. At this point, we will only take it as a fact that natural populations are genetically polymorphic. For example, the size of organisms often responds quickly to selection. Among numerous examples, Robertson and Reeve (1952) produced divergent lines differing by 25% of the initial size after 15 generations of selection in a population of *Drosophila*. One of the most obvious effects of selection in domestic animals has been to produce, e.g., dogs varying in size from that of a large rodent to that of a small calf. These examples show that ample variation exists to produce increases in variance of population characteristics as considered Section 3.5. The niche width of the population growing according to Eq. (3.72) is

$$V_t = W^2 + 2pq\varepsilon^2 , \tag{3.92}$$

where the genetical component of $V_b$ is the last term $2pq\varepsilon^2$.

The environmental component of the between phenotype variance, $V_b$, shares a lot of properties with the within phenotype component, $V_w$. In theory, this part of $V_b$ is an intrinsic property of the population which is slowly changable through time by means of natural selection. In practice, however, distinction of this component of $V_b$ is necessary mainly to allow inferences about the (in the long run) more important genetic component of $V_b$.

Models like the ones considered in Section 3.5 predict that the genetic component of $V_b$ should only be nonzero if the species packing is loose enough. This, however, is an overinterpretation of the results of such a simple model. It does, on the other hand, predict certain intuitively acceptable properties of the genetic component of $V_b$. Environmental stochasticity for example, either in general resource abundance or in unpredictable abundance of the different resource qualities, will tend to strengthen the condition for limiting similarity of species and thus allow a niche expansion of the species present. This aspect of the evolution of niche width is discussed in more detail by Roughgarden (1972, 1974a).

A change in the mean size of an animal as considered above, will effect the mean size of food particles eaten, but rarely their variance. The individual variance in particle size, i.e. the ability to select or eat particles say 2, 3, or 4 times larger than the optimum size, is a function of morphological and physiological properties of the organism with a much more complex basis than the mean size of the whole animal or of a trophic organ and will, therefore, respond much more slowly to natural selection. This property of the within phenotype component, $V_w$, is shared by the environmental component of the between phenotype component of the niche, $V_b$. Examples of different phenomena in connection with both these variances are often found by comparing different species or taxa rather than from variation within populations.

Different species may differ considerably with respect to the within phenotype component. In many taxonomic groups there are members with a strictly specialized diet and related forms using a wide variety of food items. Many estuarine animals adapt readily to a wide span of salinities and temperatures while their oceanic relatives are extremely sensitive to changes in these

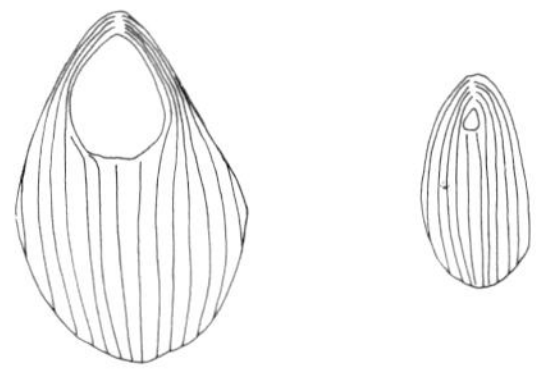

Fig. 50. The macrostome and the microstome form of the ciliate *Tetrahymena paravorax* (schematically drawn). This and some other *Tetrahymena* spp. have two alternative niches with respect to food particle size. (After Corliss, 1973)

environmental factors. There can be no doubt that the within phenotype component of the niche width is responsive to natural selection and that it changes through the evolution of the species. A most strange example of a large $V_w$ is found in the genus *Tetrahymena*. In clones of this protozoan the quality of the available food determines whether the beast will have a small or a large mouth, a micro- or a macrostome form. In cultures, and presumably in nature, populations may convert from one form to the other according to the available food and thus convert from one food niche to another (Fig. 50).

We have in this section so far discussed the niche width as a universal constant referring back to the theory in the last section. However, the width of the niche cannot be discussed without referring to the scale of measurement. The results of the theory in the last section are valid only with a scale of measurement where the utilization functions of the species in the community in question are Gaussian. With respect to food particle size selectivity, the selectivity of an animal is probably made by comparing the particle size as a fraction of the mouth, which makes it intelligible that the selected particles show a log-normal rather than a normal distribution, and that species coexisting on the basis of food particle size selectivity often form a geometric series with respect to body size. Thus, in this example the theory applies to measurements on a log-scale.

Summing up this discussion we can say (1) that the niche has a within phenotype, a between phenotype, and an age structure component, (2) that all three components have an environmental as well as a genetical basis. The genetical basis of the components is changing through the course of evolution by natural selection, and the genetical component of $V_b$ is mainly maintained by natural selection. Often the mode of the niche is more responsive to selection than the shape of the niche, as expressed by the width. (3) The total niche width is determined by interspecific competition, morphological, physiological and behavioral properties of the species and the limits of the resource spectrum—all of which tend to restrict niche width—and by intraspecific competition and environmental stochasticity which will both tend to increase niche width.

So far we have mainly had the fundamental niche in mind during this discussion although (3) above also takes into account the factors determining the niche realized by a particular population in nature. This is partly due to the fact that we have focused on resources which are exploited not occupied. The prime definition of competitive coefficients [Eq. (3.9)] applies to elements of a food resource which are eaten piece by piece; the other extreme is a habitat niche which is occupied by an individual with highest fitness, such that the competition fitness at the point $\varrho$ is proportional to max $\{0, [\mathcal{U}_i(\varrho) - \mathcal{U}_j(\varrho)]\}$. Differences in habitat niche may often be connected with limited interaction between the species solely due to spatial isolation, a phenomenon that is more thoroughly discussed in Chapter 4.

When individuals of a population do not utilize a particular resource in nature it may have two causes. The species may have evolved in the absence of that particular resource or while being excluded from it by competition, such that it lacks genotypes which form phenotypes capable of utilizing the resource. Alternatively, the individuals in the population are capable of utilizing the resource but in the studied population they cannot do so due to competition from other species. The phenomenon of *ecological release*, i.e., that a species increases its niche width in habitats devoid of competitors or in the presence of resources not available in its usual habitats need not, initially at least, have any genetical basis.

## 3.7 Evidence from Nature

The theory of the ecological niche is first of all significant in the light of observations from nature. As a predictive theory it gives insight into a number of properties of communities and populations; some examples will be given in the following.

### 3.7.1 Ecological Release and Ecological Compression

As already discussed in the previous section, populations which are normally confined to a restricted number of habitat types may in the absence of competition expand their range to other habitats. MacArthur (1972) and MacArthur and Wilson (1967) mention several examples of birds and insects which on the mainland are restricted to a single or few, often marginal habitat types. On islands with an impoverished fauna, however, the same species may be dominant in several habitats which on the mainland are inhabited by other species.

The fauna of the Baltic Sea constitutes one of the best examples of ecological release. This large brackish water sea is only about 5000 years old in its present state and no endemic species have yet evolved. This is in contrast to the much older "Pontocaspian Sea" (the Caspian Sea, the Black Sea, and the Lake Aral) in which the evolution of a spezialized brackish water fauna started already in the Miocene period (Remane and Schlieper, 1971). The Baltic Sea is, therefore, only inhabited by a relatively low number of estuarine species, a few freshwater species (mainly fish and pulmonate snails) and a few arctic marine species which are relics from the glacial period and which can endure the constantly low salinities (about 10‰ at the island of Bornholm and down to 4‰ in the Gulf of Bothnia) and the low temperatures (see Table 4). Many of the Baltic populations, however, form separate "races" which are characterized by different isozyme patterns as well as by morphological and physiological features.

The species which have a wide distribution in the Baltic are often limited to certain shallow water habitats in the inner Danish waters and the North Sea.

Table 4. An example of the reduction of species in the Baltic Sea: the species numbers of marine bivalves and prosobranch snails from the North Sea through the inner Danish waters to the inner Baltic. (Data from Remane and Schlieper, 1971)

| | North Sea | Kattegat | Øresund | Kiel Bay | Outer Baltic | Gulf of Finland |
|---|---|---|---|---|---|---|
| Bivalves | 189 | 92 | 61 | 32 | 11 | 4 |
| Prosobranchs | 210 | 101 | 69 | 26 | 13 | 2 |

The bivalve *Macoma balthica* is in the latter mentioned areas restricted to relatively clean and well sorted sand in shallow or intertidal localities. In deeper waters and in finer sediments it is substituted by its congener, *Macoma calcarea* and the related species *Abra nitida* and *Abra alba*. In the Baltic Sea, however, *Macoma balthica* is found at all depths and in all types of sediments from coarse sands to mud and silt, while the other species which require higher salinities are absent. Mud snails (*Hydrobia* spp.), which elsewhere are found only in shallow water sediments consisting of silt or fine sand, are also abundant on exposed rocks and algae in the Baltic; here the hydrobiid snails are relieved from competition by littorinid and rissoid snails which at other places are found on hard substrates. Many other examples of ecological release in the Baltic Sea may be mentioned.

We do not know to what extent genetical changes have played a role in the ecological release of the Baltic populations. Outside the Baltic, the larvae of some of the species in question show special reactions to light and to substrate quality prior to the metamorphosis which lead to habitat selection of the adult animal (see Thorson, 1964 for references). It seems likely that such mechanisms are altered or absent in the Baltic populations but this has not been investigated.

Amphipods of the genus *Gammarus* in Northern Europe constitute another example of ecological release. In most places the species *Gammarus duebeni* is restricted to marginal habitats such as rock pools and freshwater springs along the shore since this species is tolerant to changes in salinities, to extreme temperatures, and to anoxia. However, it is very rarely found in the sea or in larger freshwater localities since it is apparently an inferior competitor relative to other species of *Gammarus*, some of which are found exclusively in freshwater and others in seawater or in brackish water. On Ireland and in a few other places, however, the normal freshwater species of *Gammarus* are absent for historical reasons and here *G. duebeni* is found everywhere in freshwater (Pinkster et al., 1970).

Ecological compression means that when a species successfully colonizes a new area, the realized niche of some species already established is restricted. A well-known example is that of the arctic hare *(Lepus arcticus)* in Newfoundland. Originally it was the only hare in this region where it inhabited the tundra as well as forests. After the varying hare *(L. americanus)* was introduced, the arctic hare became restricted to the tundra (Cameron, 1958).

The limit to ecological compression of competing species within an area is ultimatively set by the principle of limiting similarity and by the probability of extinction of small populations (see also Chap. 4). Physical heterogeneity of the environment, however, is a prerequisite for habitat selection and a correlation between species numbers and environmental complexity could be expected. This has been documented for, e.g., faunas of birds in forests where more bird species are found when there are several layers of foilage (MacArthur and MacArthur, 1961) and for prosobranchs of the genus *Conus* in coral reefs (Kohn, 1968).

MacArthur (1972) and MacArthur and Wilson (1967) generalize that ecological compression dominantly takes the form of habitat compression, whereas the food spectrum of the involved populations will remain unaltered. However, since ecological compression involving resource and time niches is usually termed "character displacement," a not uncommon phenomenon also discussed by the above mentioned authors, the question is in part a semantic one. However, it may well be that successful colonizations more often lead to habitat compression than to specialization in the utilization of food resources. Food resource partitioning will probably always involve genetical changes unless it is a result of habitat selection, whereas habitat compression may at least initially simply result from mutual habitat exclusion without involving any genetical changes.

### 3.7.2 Niche Width and Species Numbers

The theoretical considerations with respect to the limiting niche overlap of coexisting species do not imply any assumptions which restrict the generality of the result. It therefore follows, that in habitats with few species either the niches are wider or they are less densely packed. The phenomenon of ecological compression and release indicate that the former explanation may often be the correct one. It may be asked how observations from nature fit the considerations of the previous section with respect to what determines the width of the niche. It was conjectured that environmental stochasticity tends to increase niche width.

The "latitudinal gradient of species diversity" is one of the most spectacular examples of differences in species richness between different areas (see Table 5). More recently it has also been found that the deep sea harbors a high number of species relative to shelf sediments (Sanders, 1968). These findings support the consideration that constant climatic conditions allow for more narrow niches while varying or unpredictable environmental conditions lead to wider niches. MacArthur (1972) mentions as an example that in a tropical climate fruits are available throughout the year and consequently some species of birds may

Table 5. The numbers of species of different taxa in three regions at different latitudes. (Data after Kormondy, 1969)

| | Labrador | Massachussetts | Florida |
|---|---|---|---|
| Coleoptera | 169 | 2000 | 4000 |
| Terrestrial snails | 25 | 100 | 250 |
| Coastal fish | 75 | 225 | 650 |
| Flowering plants | 390 | 1650 | 2500 |
| Marine bivalves | 30 | 150 | 200 |

specialize as fruit eaters. In temperate climates fruits are only available during a restricted part of the year and sedentary birds must, therefore, have wider food niches. Many examples of this sort may be given which make the latitudinal gradient in species richness intelligible and such considerations also apply to the deep sea with its constant environmental conditions.

The problem of species richness in different climates is discussed in detail by Pianka (1966), Sanders (1968), and MacArthur (1972).

### 3.7.3 Niche Overlap, Character Difference and Character Displacement

There is much evidence to show, that in accordance with the theoretical considerations, coexisting species which only differ along one niche dimension have niches which are spaced by about one standard deviation, i.e., $d/W \sim 1$ (see Figs. 40 and 51). Many other examples are given by MacArthur (1972).

It is often difficult to quantify the niche width of natural populations. Instead differences in morphological structures have been used as measures of niche overlap. Most frequently whole body sizes or the sizes of some trophic organ such as beak lengths of birds, which are believed to correlate with food particle size selection, have been used. This problem is discussed in detail by Hespenheide (1973). Morphological characters of a species may measure the "between phenotype component" ($V_b$) of niche width but the "within phenotype component" ($V_w$) must be measured seperately when data on character difference are to be analyzed in terms of niche overlap. An example of such an analysis will be given below in connection with character displacement.

Hutchinson (1959) found for a variety of animal groups that a length ratio of 1.3 allows for the coexistence of congeners based on size difference only. Schoener (1965) gave a long list of ratios of beak lengths for related birds living in the same areas. Within several taxonomic groups, ratios smaller than Hutchinson's value of 1.3 were found. However, in many cases the groups of species studied by Schoener are not really coexisting but show habitat selection within the geographical region where they are found. The fact that groups consisting of large birds (e.g., birds of prey) generally conform to Hutchinson's ratio while smaller birds (e.g., passerines) often do not, confirm this interpretation

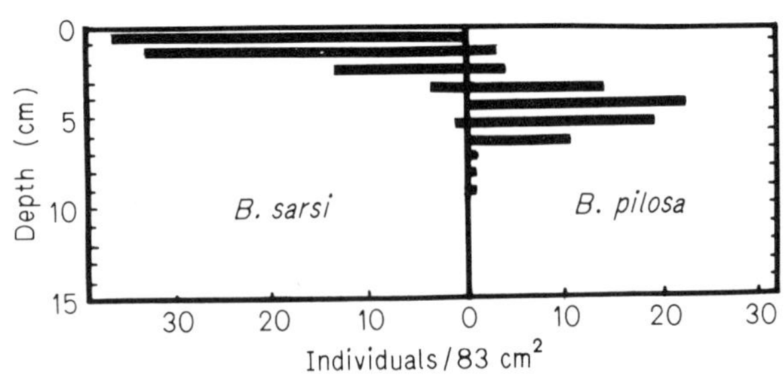


Fig. 51. The depth distribution of two sand burrowing amphipods of the genus *Bathyporeia* coexisting in a Danish beach. (Redrawn from Nicolaisen and Kanneworff, 1969)

since smaller birds more easily develop habitat selection, for example with respect to feeding places, than do larger species.

Brown and Wilson (1956) coined the term *character displacement* to describe the phenomenon, that when two closely related species coexist they may differ more from each other than allopatrically occurring populations of the same two species. Character displacement really describes two different phenomena. In zones of overlap of two related species, morphological or behavioral features which hinder hybridization may develop. This may happen when the hybrid has lower fitness than the parental forms. Character displacement may also describe changes in morphological or behavioral changes which relieve competition in zones of overlap. Here we will only discuss the latter type of character displacement.

Hutchinson (1959) reviewed the cases of character displacement known at the time; these were mainly based on metric measures of some terrestrial vertebrates and a few arthropods. Hutchinson also showed that in two corixids (aquatic hemipterans) which coexist through size difference in some localities, the larger species reproduces first so that the size ratio remains constant throughout the life cycle. Most cases of character displacement in coexisting congeners manifested themselves as a ratio in linear dimensions of about 1.3. This supports the consideration in Section 3.6, that changes in the average size of organisms may occur rapidly as a result of natural selection.

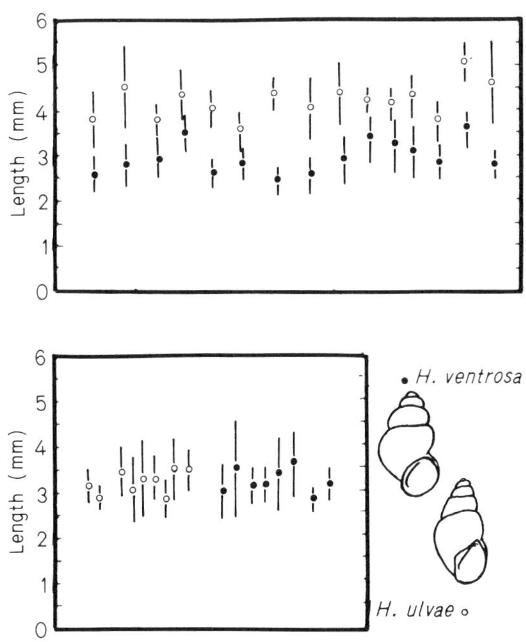


Fig. 52. The shell lengths (the mean and ±1 standard deviation) of the mud snails *Hydrobia ventrosa* and *H. ulvae* in 15 localities where they coexist (*above*) and in 17 localities where only one of the species is found (*below*). (After Fenchel, 1975b)

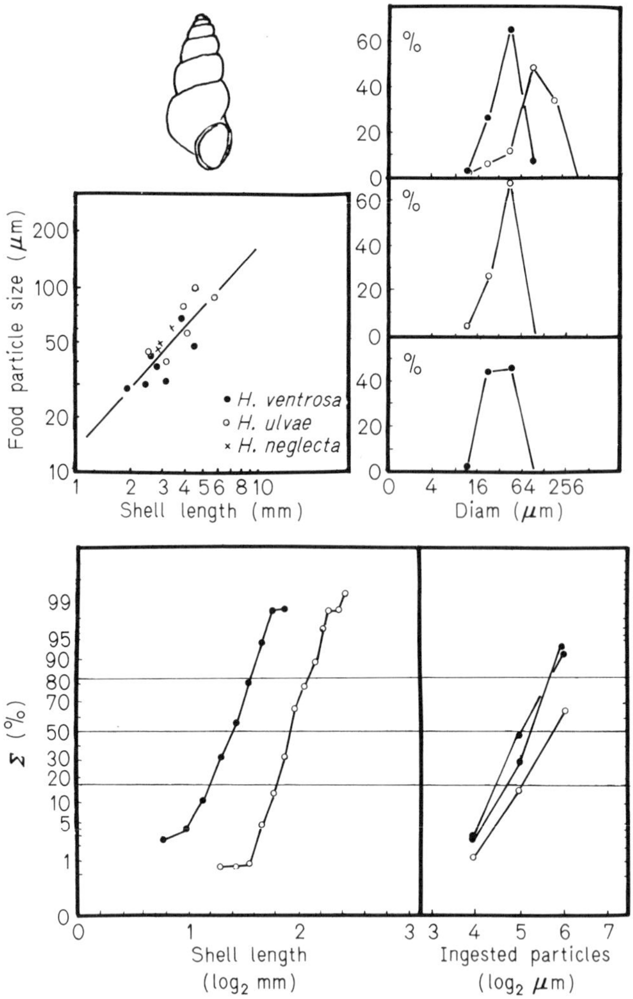


Fig. 53. *Above, left*: the average particle sizes consumed as function of shell length in the three species of *Hydrobia*. *Above right*: the range of particle sizes consumed by the total populations of *H. ulvae* and of *H. ventrosa* in a locality where they coexist (above) and where each of the species occur alone (middle and below). *Below*: the cumulative shell lengths distribution of two populations of *Hydrobia* and the cumulative particle size distribution of particles consumed by 3 individual snails plotted on probability paper. (After Fenchel, 1975b)

The three species of mud snails (*Hydrobia* spp.) living in Danish lagoons and estuaries show habitat selection with respect to salinity; the species *Hydrobia ventrosa*, *Hydrobia neglecta*, and *Hydrobia ulvae* having optimum salinities at around 20, 25, and 30‰ respectively. Although they all have wide ranges in tolerance to salinity, they usually show mutual exclusion according to salinity (and probably also some other environmental factors). In some salinity gradients and in other complex systems of lagoons, however, two or sometimes all three species may coexist. In such localities the species show character displacement with respect to size and to reproductive periods (Fenchel, 1975b). For example where *H. ventrosa* and *H. ulvae* coexist the have a size ratio of about 1.4 whereas they are of about the same size where they occur allopatrically (Fig. 52).

Hydrobiids feed on diatoms and on sediment particles, utilizing attached microflora on the latter. The average particle size ingested is a linear function of body length (Fig. 53). As also shown on the figure, the standard deviation of the ingested particles of an individual is about $0.4 \log_2$ units and the standard deviation of shell lengths of single populations is about $0.1 \log_2$ units (thus, the "within phenotype component" of niche width is larger than the "between phenotype component" + the "age structure component" in this case). In order for the species to have a niche separation of one standard deviation the size ratio would have to be $(0.4^2 + 0.1^2)^{1/2} \log_2$ units which equals a factor of about 1.3. Here we see that character displacement in fact leads to a size difference which would allow the snails to coexist and that they conform to the prediction that $d/W \approx 1$ assuming that they only differ in their resource niches with respect to food particle size (Fig. 53).

The theory in the preceding section predicts the observed character divergence in the hydrobiids, but it also predicts a decrease in the genetic component of the between individual variance, $V_b + V_a$. As seen from the above considerations the within phenotype variance, $V_w$, does indeed constitute a large part of the total variance in the food utilization. This, however, should be compared to a similar partitioning of the variance in populations where the two species do not coexist. However, the total utilization variance of the species seems not to depend on the allopatric or sympatric state (Fig. 53). Fortunately, there exist direct evidence of a decrease in the age component, $V_a$, of the between individual variance. In populations where the species live alone, the breeding season is extended and overlapping. On the other hand, in populations where the two species compete, both species tend to have a restricted and nonoverlapping breeding period (Fenchel, 1975b).

# 4. Geographically-Structured Populations and Migration-Extinction Equilibria

In the two previous chapters we have primarily considered population interactions under the tacit assumption of homogeneous environments and it was possible to arrive at theoretical conclusions which give insight in many observations from nature. Yet, nature seldom consists of homogeneous environments but must often be described as a heterogeneous mosaic of more or less isolated habitat patches each harboring larger or smaller populations, with migration between them. This structure has to be taken into account when a complete description of ecosystems is attempted. Such considerations lead to conclusions which in part differ qualitatively from those arrived at in the previous chapters. However, there is no contradiction between the two ways of looking at biological communities. Rather it is a question of scale; within each habitat patch the mechanisms outlined in the previous chapters are believed to function. But the interaction between a larger number of more or less isolated environmental patches interconnected by low rates of migration will result in systems with special, and in many ways, complex features.

Hutchinson (1951) and Skellam (1951) were the first to consider that more species, depending on the same resource, could coexist within a patchy environment, given certain rates of local extinctions and certain migration rates between patches.

Consider say a beech forest in which the trees are always able to exclude other plants by competition. We would expect such an area to remain a homogeneous and pure stand of trees devoid of other plant species. However, if patches are cleared at intervals due to, e.g., fires or timbering, various annual plants and shrubs could colonize these patches since they grow and reproduce faster than trees. Eventually the trees would take over the patch again by the competitive exclusion of the other plants. However, if the annual plants in the mean time had produced seeds and these had been dispersed to new, cleared patches then the plants could subsist in the forest by their ability for dispersal and fast growth, provided that local clearings appear sufficiently often. Such species were termed *fugitive species* by Hutchinson but they are often referred to as *opportunistic species* or simply as "weeds". The considerations clearly apply to species depending on emphemeral habitats or resources, and to environments characterized by periodical disturbances such as temporal ponds, carrion, dung, etc. Several species may well utilize such resources through what is known as a succession. The species which are the first to utilize the resources and later are to

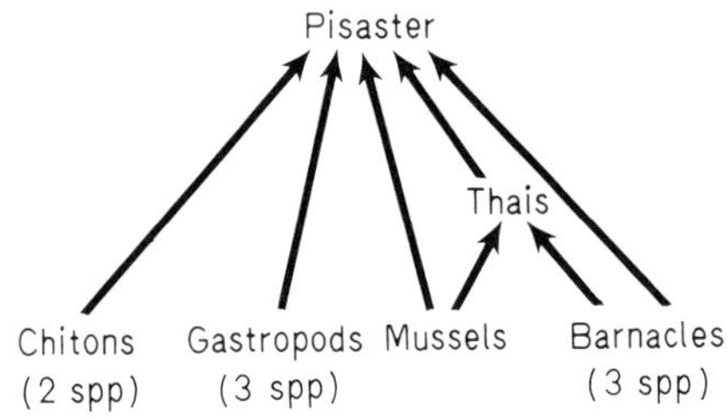


Fig. 54. The food web of a rocky shore on the Washington coast. (Redrawn from Paine, 1966)

become excluded by other species, are superior with respect to reproductive potential and dispersal relative to the climax species. The study of Paine (1966) also illustrates the coexistence of competitors in a fugitive equilibrium. The intertidal rocks along the Pacific coast of Washington harbor a fauna which forms a food web as shown in Figure 54. When Paine removed the "top predator", the seastar *Piaster*, within an area, then the numbers of the remaining species dropped and eventually mussels totally dominated the rocks. This experiment has been used for exemplifying the role of predators in stabilizing the coexistence of competing prey species. The correct interpretation is probably that the seastar clears patches on the rocks which are then colonized by a variety of species. The number of species in a patch is eventually reduced by competitive exclusion leaving only mussels. By that time, however, the seaster has cleared new patches which are open for colonization.

In the following we will investigate some models which describe aspects of coexistence in heterogeneous environments with migration and local extinctions. But first we will analyze the effect of migration on a single population and explore the effect of competition in a community with geographical subdivision and limited mixing. These models correspond to the models applied in population genetics (Wright, 1969).

## 4.1 Migration Pressure and Competition

We will here discuss a model described by Fenchel (1975a); it was inspired by the distribution patterns of mud snails in Danish estuaries and lagoons. As already discussed, the three species of *Hydrobia* have different optima with respect to salinity. The species which prefers the highest salinities forms large, widely distributed populations in the sediments of the more open waters, whereas the two brackish water species are mainly confined to small isolated populations in coves, lagoons and the innermost parts of estuaries where the salinities are low. All three species, however, have wide ranges of tolerance with respect to salinity and could probably in the absence of the two other species sustain populations over nearly the total environmental range populated by the genus. The question here is what will be the species composition of a small patch, $H_1$, in which species $x_1$ is a superior competitor will be, if $H_1$ is surrounded by a large large population of species $x_2$, from which individuals migrate into $H_1$ at a constant rate $m$ (Fig. 55).

We will look at the growth of $x_1$ and $x_2$ in $H_1$ and assume it is described by the modified Lotka-Volterra equations [cf. Eq. (2.12)]:

$$\begin{aligned} dx_1/dt &= x_1 r_1(K_1 - x_1 - \beta_{12}x_2)/K_1 \\ dx_2/dt &= x_2 r_2(K_2 - x_2 - \beta_{21}x_1)/K_2 + m\,, \end{aligned} \tag{4.1}$$

where $m$ ($\geqq 0$) is the migration rate of species 2 from $H_2$ into $H_1$. According to the assumptions both $K$s and $r$s are positive. The interesting situations are those in which for $m=0$ species 1 excludes species 2 under all or a part of the initial conditions. This is the case if

$$K_1 > K_2/\beta_{21}\,, \tag{4.2}$$

which will be assumed in the sequel.

The assymetry of the model is an approximation because emigration to $H_2$ of species 1 will tend to modify the composition of the immigrants to $H_1$. However, this may be neglected as long as $H_1$ is very small compared to $H_2$.

The possible equilibria of Eq. (4.1) are found by equating the right sides to zero and solving for $x_1$ and $x_2$. One solution, the trivial one, where species 1 is absent from $H_1$, is

$$\hat{x}_{1t} = 0 \quad \text{and} \quad \hat{x}_{2t} = K_2/2 + (K_2^2/4 + mK_2/r_2)^{1/2}\,. \tag{4.3}$$

This equilibrium is unstable provided $\hat{x}_{2t} < K_1/\beta_{12}$, i.e., if

$$m < r_2 K_1(K_1 - \beta_{12}K_2)/(K_2\beta_{12}^2)\,. \tag{4.4}$$

Thus if $x_1 > 0$ initially, coexistence will occur if Eq. (4.4) is fulfilled.

The coexistence equilibrium depends on the type of competition. In case of exploitative interaction with incomplete niche overlap, i.e., when

$$\beta_{12} < 1/\beta_{21}\,,$$

there may be a single coexistence equilibrium given by

$$\hat{x}_2 = A + (A^2 + mB)^{1/2} \quad \text{and} \quad \hat{x}_1 = K_1 - \beta_{12}\hat{x}_2\,, \tag{4.5}$$

where

$$\begin{aligned} A &= (K_2 - \beta_{21}K_1)/[2(1 - \beta_{12}\beta_{21})] < 0\,, \\ B &= K_2/[r_2(1 - \beta_{12}\beta_{21})] > 0\,. \end{aligned} \tag{4.6}$$

This equilibrium exists and is stable if $\hat{x}_1 > 0$, which is equivalent to condition Eq. (4.4). If Eq. (4.4) is not fulfilled, then species 2 will exclude its superior competitor, species 1, from $H_1$. The same conclusions can be drawn if $\beta_{12} = 1/\beta_{21}$.

With antagonistic interaction, i.e., when $\beta_{12} > 1/\beta_{21}$, there may be two coexistence equilibria,

$$\begin{aligned} \hat{x}_{2s} &= A - (A^2 + mB)^{1/2} \quad \text{and} \quad \hat{x}_{1s} = K_1 - \beta_{12}\hat{x}_{2s} \quad \text{and} \\ \hat{x}_{2u} &= A + (A^2 + mB)^{1/2} \quad \text{and} \quad \hat{x}_{1u} = K_1 - \beta_{12}\hat{x}_{2u}, \end{aligned} \tag{4.7}$$

where $A$ and $B$ are given by Eq. (4.6), but now $A > 0$ and $B < 0$.

The equilibrium $(\hat{x}_{1s}, \hat{x}_{2s})$ is always stable when it exists, and the equilibrium $(\hat{x}_{1u}, \hat{x}_{2u})$ is always unstable. When condition Eq. (4.4) is fulfilled, i.e., when the $(0, \hat{x}_{2t})$ equilibrium is unstable, the $s$-equilibrium exists and the $u$-equilibrium does not exist. On the other hand, if Eq. (4.4) is not fulfilled, the equilibria Eq. (4.7) may still be relevant. If

$$\beta_{12} \leqq 2K_1/(K_1\beta_{21} + K_2), \tag{4.8}$$

then neither of the equilibria (4.7) exists when Eq. (4.4) is not fulfilled and $(0, \hat{x}_{2t})$ is the only stable equilibrium. Thus, the situation is the same as in the case $\beta_{12} \leqq 1/\beta_{21}$. However, if Eq. (4.8) is false, then for $m$ belonging to the interval

$$r_2K_1(K_1 - \beta_{12}K_2)/(K_2\beta_{12}^2) < m < r_2(K_2 - \beta_{21}K_1)^2/[4K_2(\beta_{12}\beta_{21} - 1)] \tag{4.9}$$

both equilibria of Eq. (4.7) exist, so that there are two stable equilibria $(\hat{x}_{1s}, \hat{x}_{2s})$ and $(0, \hat{x}_{2t})$ and one unstable equilibrium separating them on the curve

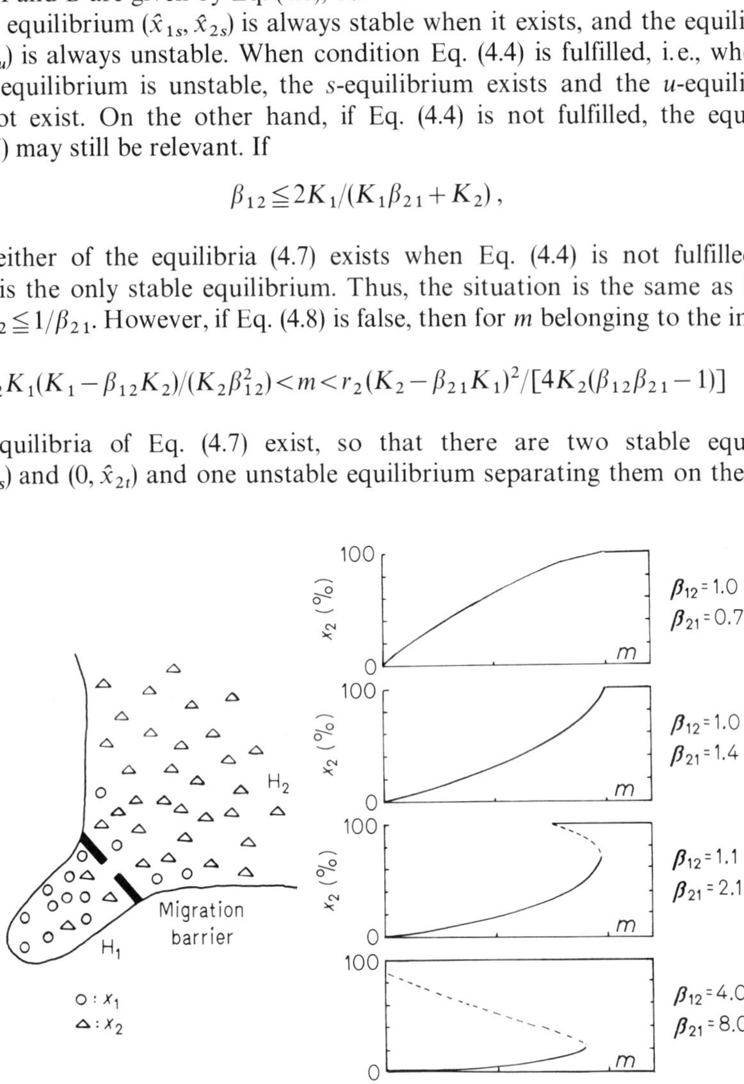


Fig. 55. A system such as described by the model [Eq. (4.1)]. Species 1 is a superior competitor in habitat 1 but is subject to a certain migration of species 2 from habitat 2 (*left*). *Right*: equilibrium composition of the competing species in habitat 1 as function of $m$ with all other constants fixed ($K_1 = 2K_2$) for four different sets of values of competition coefficients. *Broken lines*: unstable equilibria. (Altered after Fenchel, 1975a)

determined by $dx_2/dt=0$. If $\beta_{12}>K_1/K_2$, the left side of (4.9) [right side of Eq. (4.4)] is negative and has to be replaced by 0, signifying that the equilibrium $(0, \hat{x}_{2t})$ is stable for all $m$. In spite of this, species 1 can sustain a population in $H_1$ as long as it is sufficiently numerous. In practice, a disturbance of the population through a discrete, exceptionally high migration rate may cause a sudden change from one stable equilibrium to the other resulting in a permanent loss of species 1 in the area. If $m$ is greater than or equal to the right side of Eq. (4.9), neither of the equilibria of Eq. (4.7) exists, so species 2 will exclude species 1.

Figure 55 shows the equilibrium compositions $\hat{x}_2/(\hat{x}_1+\hat{x}_2)\%$ in the various cases as functions of $m$ with all other constants fixed.

The model gives a number of qualitative predictions of interest. Figure 55 and the Eq. (4.4) show that a superior competitor in a small habitat (a small $K_1$) may be excluded by an inferior competitor if the latter has a high rate of migration into the habitat (i.e., large populations in the neighboring habitats). This prediction which can be confirmed in the field (Fig. 56 and Fenchel, 1975a) is also of significance for the theory of island biogeography (see Sect. 4.5).

Given moderate migration rates or habitats of about equal sizes we may expect coexistence of the species in both habitats although one of the species would always be excluded in a homogeneous area.

If the migration rates decrease regularly when going from the center of one habitat towards the center of the other, then $m$ on Figure 55 may be taken as a measure of distance and the curves will describe the gradual transition from one species to the other. The steepness and exact position of the slope with respect to the factors which determine the outcome of competition will also depend on migration rates which again depend on the presence or absence of migration

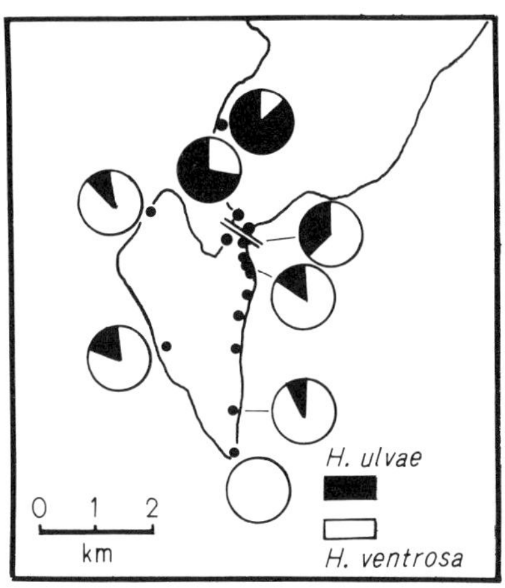


Fig. 56. The distribution pattern of two species of *Hydrobia* inside and outside a brackish water lagoon in the Limfjord, Denmark. The situation is parallel to Figure 55. Outside the lagoon *H. ulvae* is superior and inside *H. ventrosa* is superior as competitor but a gradient in the composition of the snail fauna is found. (Redrawn from Fenchel, 1975a)

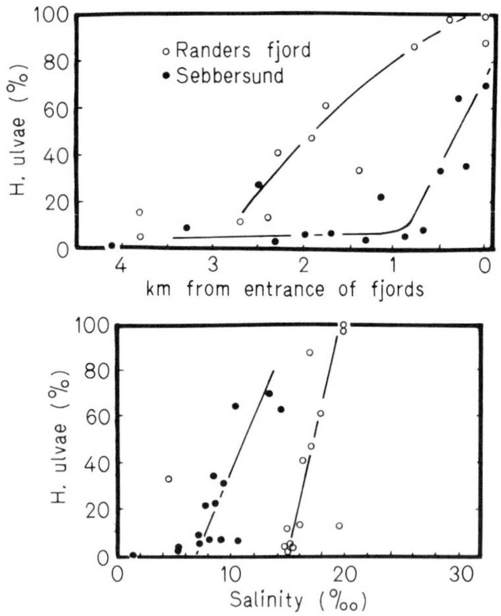


Fig. 57. The transition from *H. ulvae* to *H. ventrosa* as the dominating *Hydrobia* sp. in two fjords with salinity gradients as function of distance from the inlet and as function of salinity. *H. ulvae* extends its distribution towards lower salinities where the migration distance is the shortest. (After Fenchel, 1975a)

barriers and the sizes and densities of the populations. This is exemplified on Figures 56 and 57. It shows the gradient between *H. ulvae* (the marine species, corresponding to species $x_2$ of the model) and the brackish water species *H. ventrosa* in two estuaries with a salinity gradient. It can be seen that in the shortest gradient *H. ulvae* penetrates into lower salinities in accordance with the model.

## 4.2 Migration Between Identical Habitat Patches

Levin (1974) studied a model of a system with two identical patches and with two competing species with antagonistic interrelationship; i.e., in a homogeneous patch one of the species would always be excluded according to the initial population sizes.

In the notation of Eq. (2.28) we have

$$dx_1/dt = x_1(1 - x_1 - \alpha x_2),$$
$$dx_2/dt = x_2(1 - x_2 - \alpha x_1), \qquad (4.10)$$

where $\alpha_{12} = \alpha_{21} = \alpha > 1$.

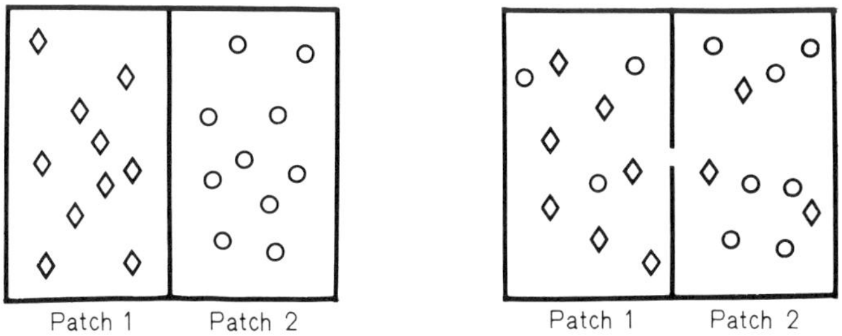


Fig. 58. A graphical representation of the model [Eq. (4.12)]

The system [Eq. (4.10)] has 4 equilibria;

$$\begin{aligned} &1.\ \hat{x}_1=0;\hat{x}_2=0\\ &2.\ \hat{x}_1=0;\hat{x}_2=1\\ &3.\ \hat{x}_1=1;\hat{x}_2=0\\ &4.\ \hat{x}_1=\alpha/(1+\alpha);\hat{x}_2=\alpha/(1+\alpha)\,, \end{aligned} \tag{4.11}$$

of which 2. and 3. are stable (cf. Sect. 2.2).

Now consider a system consisting of two isolated patches (Fig. 58). If this system is inhabited by $x_1$ and $x_2$ we have according to Eq. (4.11), 16 possible equilibria of which the following 4 are stable;

$$\begin{aligned} &[\hat{x}_1=1;\hat{x}_2=0]\,[\hat{x}_1=0;\hat{x}_2=1];\,[\hat{x}_1=0;\hat{x}_2=1]\,[\hat{x}_1=1;\hat{x}_2=0]\\ &[\hat{x}_1=0;\hat{x}_2=1]\,[\hat{x}_1=0;\hat{x}_2=1];\,[\hat{x}_1=1;\hat{x}_2=0]\,[\hat{x}_1=1;\hat{x}_2=0]\,, \end{aligned}$$

where each pair of bracket describes the conditions in patches 1 and 2. These situations could initially be brought about by historical or stochastic events and they will remain stable since there is no interaction between the two patches. Now consider one of the equilibria in which different species have become established in the two patches, e.g., that species 1 is established in patch 1 and species 2 is established in patch 2. Let us now allow for a low migration rate of individuals between the patches. For patch 1 Eq. (4.10) now becomes

$$\begin{aligned} dx_1^1/dt&=x_1^1(1-x_1^1-\alpha x_2^1)+m(x_1^2-x_1^1)\,,\\ dx_2^1/dt&=x_2^1(1-x_2^1-\alpha x_2^1)+m(x_2^2-x_2^1)\,, \end{aligned} \tag{4.12}$$

where $x^1$ and $x^2$ denote the population sizes in patches 1 and 2 respectively, and where $m$ is a migration constant which is considered identical for the two species such that migration is assumed to be proportional to the population sizes.

Now it can be seen intuitively, that as long as $m$ is small the system will remain stable (i.e., both species will persist) since in patch 1 the system will tend toward the ($\hat{x}_1 = 1$, $\hat{x}_2 = 0$) equilibrium and in order to change this trend a minimum influx of species 2 is necessary. So for sufficiently small values of $m$ we will have coexistence in both patches. Levin (1974) proved this result formally, determined the new equilibrium as a function of $m$, and showed that the condition for coexistence is that

$$0 < m < (1/2)(\alpha - 1)/(\alpha + 1) \tag{4.13}$$

If the right inequality of Eq. (4.13) is not satisfied one of the two species will take over the entire system [a violation of Eq. (4.13) will lead to an unstable equilibrium in (4.12)].

This result is a special case of the general mathematical theory of weak interactions (Karlin and McGregor, 1972). This theory as well as the model discussed above show that the heterogeneity constituted by a migration barrier between identical patches is sufficient to maintain a higher degree of complexity than a totally homogeneous environment would allow.

The principle may be extended to $n$ identical patches with a limited rate of migration between them. Such a system may in principle maintain $n$ otherwise mutually exclusive species in a stable coexistence.

It is clear that the equilibria discussed here are in a sense neutral, i.e., if one species population due to say stochastic fluctuations becomes rare it may go extinct altogether, since the system is not globally stable; it can, therefore, be expected that complex systems maintained as discussed above will eventually simplify to a lower number of species. However, such a loss of species may be a very slow process. It can, therefore, be expected that in nature we may find systems of identical patches (islands, small isolated forests, lakes, mountain tops) with different faunal compositions, which were originally established by some historical event, and which will seem stable as may be inferred from the observation of some migration between patches.

## 4.3 Stochastic Extinction of Populations

In the previous chapters we have confined our interest to situations where the population sizes are virtually infinite i.e., we have assumed populations so large that they can be described by a continuous variable. In this framework, the population of a species may go extinct either because the environment cannot support it or because of the influence of other species. Extinction in this context means that the population size vanishes to zero. In order to describe this process strictly, we have to realize that at some stage our description of population size as a continuous variable is invalid; we will have to describe the population as a collection of individuals for whom death and reproduction are discrete, stochastic events, i.e., we have to take into account what May (1973a) termed *demographic stochasticity*. Similar reservations apply to conditions for the invasion of a species into a habitat. A newly introduced species will be represented by few individuals and demographic stochasticity will be decisive for the success of the colonization. This is especially important for the description of a single immigration event;

in the deterministic theory we have so far explored, the condition for potential success of invasions assumes recurrent immigration events.

The simplest example of a population growth model where the finite number of individuals is taken into account is a discrete generation model described by the Galton-Watson or branching process. Assume that we have a population size $X_n$ at generation $n$ and that the $i$'th individual, $i=1,2,\ldots,X_n$, gets $P_i$ progeny, then the next generation will have the size

$$X_{n+1}=\sum_{i=1}^{X_n} P_i . \tag{4.14}$$

Now, assume that the individuals reproduce independently and that the actual number of offspring $P_i$ of an individual is drawn from a common probability distribution:

$$\text{Probability}(P_i=k)=P(P_i=k)=p_k , \tag{4.15}$$

for $k=0,1,2,\ldots$, where

$$\sum_{k=0}^{\infty} p_k=1 . \tag{4.16}$$

The event $P_i=0$ means that the individual $i$ dies before reproduction. The event $P_i=1$ either means that the individual dies after leaving one offspring or that the individual survives without leaving offspring. However, in the following we will always interpret the process in terms of, e.g., an annual organism with a discrete breeding season; i.e., all individuals die after the breeding season and may or may not leave progeny.

With these assumptions the stochastic process $X_0, X_1, X_2,\ldots,X_n,\ldots$, is called a branching process with offspring distribution [Eq. (4.15)]. The model is a discrete time and discrete number analog of the unlimited growth model [Eq. (1.1)] and we may expect it to have similar properties. However, we are only interested in the stochastic description for low population sizes, it may also be a good approximation to the initial growth of the logistic model, if the carrying capacity is sufficiently large.

An introduction to the properties of this kind of process is found in Karlin (1966); more advanced results are given by Harris (1963). Here we will give some fundamental properties of the model.

If we start a population with one individual, i.e., $X_0=1$, then the probability of dying out at the first breeding period is $q_1=p_0$. If the initial individual leaves $k$ offspring, $X_1=k$, then the probability of extinction by the next generation is $p_0^k$. The event of having $k$ offspring has the probability $p_k$ so that the probability of having $k$ offspring in the first generation and then dying out in the second generation is $p_k p_0^k$. Summing all the probabilities of having a zero population after the two first breeding periods we get the total probability of extinction at time 2, i.e., $X_2=0$, as

$$q_2=q_1+\sum_{k=1}^{\infty} p_k p_0^k . \tag{4.17}$$

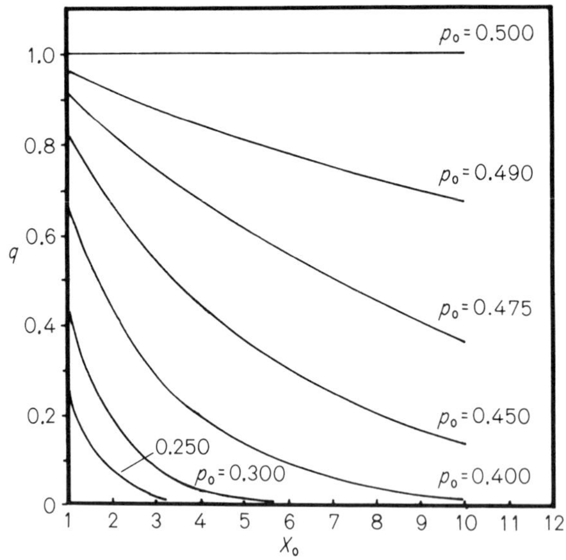


Fig. 59. The probability of extinction for populations of an organism with binary fission and with different probabilities of dying before dividing ($p_0$) as function of the founding population size ($X_0$)

This argument may be continued so that if the probability of $X_n = 0$ is $q_n$, then

$$q_{n+1} = q_n + \sum_{k=1}^{\infty} P(X_n = k) p_0^k . \tag{4.18}$$

Thus the probabilities of zero population sizes form an increasing sequence $p_0 = q_1 < q_2 < \ldots < q_n < \ldots$ converging to the probability $q$ which is within the bounds

$$p_0 < q \leqq 1 . \tag{4.19}$$

The extinction probability $q$ is unity if and only if the mean offspring size $m$ is less than or equal to one, where

$$m = \sum_{k=1}^{\infty} k p_k . \tag{4.20}$$

If $m > 1$, then the extinction probability is strictly less than unity and with the probability $1 - q$ the population size will grow to infinity. The expected value of the population size in the $n$'th generation is $m^n$, this may be seen from the definition [Eq. (4.14)].

The extinction probability $q$ pertains to a population starting with one individual, i.e., $X_0 = 1$, but if we start out with $x_0$ individuals, then the descendants of each of the initial individuals grow independently as populations started from one individual. The probability of extinction is therefore, $q^{x_0}$ (Fig. 59). By a similar argument as the one leading to Eq. (4.17) we therefore get for $q$;

$$q = \sum_{k=0}^{\infty} p_k q^k . \tag{4.21}$$

The extinction probability $q$ is then the smallest positive root of this equation.

With respect to density dependent models, such as the logistic models described in Eq. (1.9) and Eq. (1.13), no stochastic analog has been formulated or analyzed sufficiently. However, we can get a lower limit for the extinction probability for this model. In the branching model for unlimited growth with $m > 1$, the population either dies out with a probability $q$ or it keeps growing away from the extinction state. In the density dependent model, if the population does not die out initially, it will have a size $X_n \approx K$ and with an offspring mean of one. Thus a population size of $K$ will have an extinction probability which is greater than a branching process with a mean $m > 1$ started at $X_0 = K$ which will have the extinction probability $q^K$. This must be a gross underestimate of the extinction probability of the density dependent population since if it grows to a size above $K$, it is expected to return to $K$ again. Therefore, the extinction probability will approach unity but the expected time for the extinction will increase with increasing $K$. This may be formulated as follows: Given a large collection of populations of the size $K$, then these populations will go extinct at a rate which increases with decreasing $K$.

Stochastic variation in population size may also be due to *environmental stochasticity*, e.g., a variation in resource abundance or in a density independent death rate. Let us first look at the simplest model of unlimited growth with periodically varying multiplication rates,

$$x' = (R+1)x \tag{4.22}$$

such that $R$ varies from generation to generation according to the scheme: $R_1, R_2, \ldots, R_k, R_1, R_2, \ldots, R_k, \ldots$ The fate of the population may, therefore, be expressed at the $k$'th generation by

$$x^{(k)} = (R_1+1)\,(R_2+1)\ldots(R_k+1)\,x\,, \tag{4.23}$$

revealing that the population is growing if the product $(R_1+1)\,(R_2+1)\ldots(R_k+1)$ is larger than one and that it will vanish if the product is less than unity. This may be expressed in terms of the mean,

$$(1/k) \sum_{j=1}^{k} \log(R_j+1) \tag{4.24}$$

which determines whether the population grows or vanishes according to whether the mean is larger or smaller than zero.

Now turning to the stochastically varying $R$, we pick for each generation a value for $R$ according to the probability distribution of $R$. The environmental variation is, therefore, described by the sequence of independent random variables $R_1, R_2, \ldots, R_k, R_{k+1}, \ldots$, and the fate of the population at any generation $k$ is determined by equations like Eq. (4.23). From Eq. (4.23) we immediately get the population mean after $k$ generations as

$$EX^{(k)} = [E(R+1)]^k x\,, \tag{4.25}$$

i.e., the mean population size increases or decreases according to whether the expected value $E(R+1)$ is greater or less than unity. However, the expected value of the population size may not be very informative with respect to the form of the distribution of the population size (Lewontin and Cohen, 1969). The mean may be high because the population has a vanishing small probability of growing extremely fast while it at the same time has a high probability of going extinct. A more interesting quantity is the probability that the population after $k$ generations is above some size $y$, i.e., the quantity

$$\text{Prob.}(X^{(k)} > y)\,. \tag{4.26}$$

In order to find this probability we can use the idea expressed in Eq. (4.24);

$$\begin{aligned}\text{Prob.}(X^{(k)} > y) &= \text{Prob.}[\log(X^{(k)}) > \log(y)] \\ &= \text{Prob.}\left[\sum_{j=1}^{k} \log(R_j+1) > \log(y) - \log(x)\right],\end{aligned} \tag{4.27}$$

where we have used Eq. (4.23). We may rewrite Eq. (4.27) as

$$\text{Prob.}(X^{(k)} > y) = \text{Prob.}\left\{(1/k)\sum_{j=1}^{k} \log(R_j+1) > [\log(y) - \log(x)]/k\right\}. \tag{4.28}$$

For large values of $k$ the arithmetic mean on the left side of the inequality will be very close to the expected value $E[\log(R+1)]$ of the logarithm of the growth factor with a probability approaching unity (from the law of large numbers). On the other hand, the quantity $[\log(y/x)]/k$ on the right side of the inequality becomes vanishing small for large values of $k$. Hence, we can conclude that the probability that the population number $X^{(k)}$ is larger than a constant $y$ for $k$ (the number of generations) tending to infinity is positive if and only if the expected logarithmic growth rate $E[\log(R+1)]$ is larger than zero. When this is the case the probability [Eq. (4.28)] in fact approaches unity which may be seen using the central limit theorem (Lewontin and Cohen, 1969). Comparison with the behavior of the mean population size is obtained by observing that the logarithm is a concave function such that by Jensen's inequality we get

$$\log[E(R+1)] > E[\log(R+1)]\,; \tag{4.29}$$

i.e., the growth of the population guarantees increase of the mean but the opposite cannot be concluded.

This result generalizes to the branching process model where the condition for positive probability for growth is that $E[\log(m)]$ is positive; in this case the growth of the population will be approximately given by

$$X_n \approx \exp\{nE[\log(m)] + \sqrt{n}\, V\},$$

provided that it does not go extinct, and where $V$ is the variance over the random environment of $\log m$, the logarithm of the mean offspring number in a particular generation (Keiding, 1975).

May (1973b) suggested a model for investigating the influence of a randomly varying environment on density dependent growth. In this model the population grows according to the differential equation

$$dx/dt = x(A - x), \tag{4.30}$$

where $A$ is given by the formal equation

$$A = A_0 + w(t). \tag{4.31}$$

In Eq. (4.31) $w$ is a white noise with the mean zero and the variance $s^2$, so that $A$ becomes a white noise with the mean $A_0$. (A white noise is a formal expression for the derivative of the stochastic process of Brownian motion, such that a Brownian motion with variance $s^2$ is the solution to the differential equation $dB/dt = w(t)$.) The integration of a stochastic differential equation like Eq. (4.30) involves some ambiguities (see Feldman and Roughgarden, 1975) depending on the applied integral calculus. May (1973b) got by one integral calculus the condition

$$s^2/2 < A_0 \tag{4.32}$$

for growth of the population. Feldman and Roughgarden (1975) using an alternative approach got no condition on the size of the variance. Furthermore, Feldman and Roughgarden (1975) point out the difficulty in the biological interpretation of the quantity $A$, because in the model [Eq. (4.30)] it is the growth rate of $x$ for $x$ close to zero, but at the same time it plays a role as the equilibrium population size in the deterministic model. Feldman and Roughgarden (1975) suggest and analyze an alternative model where

$$dx/dt = rx\{1 - x[1 + w(t)]/K\}. \tag{4.33}$$

Here the random variation is put on the carrying capacity, or rather the inverse carrying capacity, and the population can grow and exist independent of the variance of $w$.

If we return to the discrete formulation of population growth, i.e., either Eq. (1.10) or Eq. (1.14) then for very small population sizes the population growth is approximately described by Eq. (4.22) such that the condition for the successful invasion of a species into a habitat with a randomly varying environment is

$$E[\log(1 + R)] > 0. \tag{4.34}$$

If the growth rate $R$ is small, then we may approximate Eq. (4.34) by

$$\begin{aligned} E[\log(1+R)] &= E(R-1/2R^2+\ldots) \\ &\approx E(R)-1/2\mathrm{Var}(R)+E(R)^2 \\ &\approx E(R)-1/2\mathrm{Var}(R)\,, \end{aligned} \tag{4.35}$$

where Var signifies the variance. Thus, Eq. (4.34) is fulfilled and the population size grows when small if the mean growth rate is larger than half the variance of the growth rate (Keiding, 1975). When the growth rate is small, then Eq. (4.22) may be viewed as an approximation to a continuous time growth model (full details are given in Keiding, 1975), such that May's (1973b) condition seems to resemble the condition for the initial growth of a species in a randomly varying environment.

## 4.4 Fugitive Equilibria

We will now return to the problem of fugitive equilibria in a patchy environment with local extinctions and recolonization through migration between patches, such as the beech wood example discussed in the introduction to this chapter. Recently, several theoretical studies on this topic have been carried out (Cohen, 1970; Levins and Culver, 1971; Horn and MacArthur, 1972; Levin and Paine, 1974; Slatkin, 1974). In all these models the total population size of a species is measured by the number of patches inhabited by the species. No account of the dynamics within each is kept, only the events of extinction or colonization of a species in a patch are considered. All models are based on simplifying assumptions and differ in some important aspects (Slatkin, 1974). The main conclusion of all these models, however, is that under wide conditions it is possible for species to coexist in a patchy environment even if the same species cannot coexist in a homogeneous population.

Consider an area consisting of many isolated and identical patches which may be inhabited by two competing species. Let $P_1$ and $P_2$ be the frequencies of patches which harbor species 1 and 2 respectively and let $P_{12}$ be the frequency of patches with both species present such that $P_0=1-P_1-P_2-P_{12}$ is the frequency of empty patches. The population size of species $i$ is measured by

$$Q_i=P_i+w_iP_{12}\,, \tag{4.36}$$

where the weighting factor $w_i$ ($\leqq 1$) describes that the species $i$ may on the average sustain a smaller population in a patch where its competitor is present. The invasion and colonization of an empty patch is assumed to occur with the rate $m_iQ_iP_0$, i.e., proportionally to the abundance of the species and to the number of empty patches. The populations of species $i$ are assumed to go extinct with the rate $e_i$ in patches were it occurs alone so that the rate of production of empty patches by species $i$ is $e_iP_i$. If we now, in order to get started, assume that a species cannot invade a patch already inhabited, i.e., $P_{12}=0$, the system may be described by the following equations;

$$\begin{aligned} dP_0/dt &= -(m_1Q_1+m_2Q_2)\,P_0+e_1P_1+e_2P_2\,, \\ dP_1/dt &= m_1Q_1P_0-e_1P_1 \quad [=P_1(m_1P_0-e_1)]\,, \\ dP_2/dt &= m_2Q_2P_0-e_2P_2 \quad [=P_2(m_2P_0-e_2)]\,, \end{aligned} \tag{4.37}$$

where from Eq. (4.36), $Q_i=P_i$. If we start out with an empty area, i.e., $P_0=1$, then $i$ can invade if $m_i>e_i$ and the frequency of inhabited patches will increase and stabilize at

$$\hat{P}_i=1-\hat{P}_0=1-e_i/m_i\,. \tag{4.38}$$

These equilibria for $i=1,2$ are the only equilibria of the system [Eq. (4.37)], of which one is stable and one is unstable. This is seen by considering a situation where species 1 is alone in the area and observe that the condition for species 2 to be able to invade, viz. that $m_2\hat{P}_0-e_2>0$, from Eq. (4.38) is equivalent to

$$e_2/m_2<e_1/m_1\,. \tag{4.39}$$

Thus species 2 can invade if it is able to sustain a higher occupation frequency than species 1. If it invades it will increase to the frequency given by Eq. (4.38) and thus exclude species 1.

This situation changes radically if we allow for the possibility of cohabitation of a patch. Let species $i$ invade and inhabit a patch already occupied by the other species with a rate proportional to $(m_i-\mu_i)$, where $\mu_i$ typically is positive meaning that the presence of the other species decreases the chance of a successful colonization. Further let species $i$ go extinct in cohabited patches with the rate $(e_i+\varepsilon_i)$ where $\varepsilon_i$ is typically positive due to the lower population sizes which are the consequence of competition. Building these features into the system [Eq. (4.37)] yields the following equations,

$$\begin{aligned} dP_0/dt &= -(m_1Q_1+m_2Q_2)\,P_0+e_1P_1+e_2P_2\,, \\ dP_1/dt &= m_1Q_1P_0-e_1P_1-(m_2-\mu_2)\,Q_2P_1+(e_2+\varepsilon_2)\,P_{12}\,, \\ dP_2/dt &= m_2Q_2P_0-e_2P_2-(m_1-\mu_1)\,Q_1P_2+(e_1+\varepsilon_1)\,P_{12}\,, \\ dP_{12}/dt &= (m_1-\mu_1)\,Q_1P_2+(m_2-\mu_2)\,Q_2P_1-(e_1+\varepsilon_1+e_2+\varepsilon_2)\,P_{12}\,, \end{aligned} \tag{4.40}$$

(Slatkin, 1974). If the area only contains one species, then the equilibrium frequency of inhabited patches is given by Eq. (4.38). Now, let species 1 prevail in the area and introduce species 2 in low numbers. The fate of this species is then approximately described by the linear system (neglecting the squares and the products of $P_2$ and $P_{12}$)

$$\begin{aligned} dP_2/dt &= a_{11}P_2+a_{12}P_{12}\,, \\ dP_{12}/dt &= a_{21}P_2+a_{22}P_{12}\,, \end{aligned} \tag{4.41}$$

where the matrix **A** is given by

$$\begin{aligned} a_{11} &= m_2\hat{P}_0 - e_2 - (m_1-\mu_1)\,\hat{P}_1\,, \\ a_{12} &= m_2 w_2 \hat{P}_0 + (e_1+\varepsilon_1)\,, \\ a_{21} &= [(m_1-\mu_1)+(m_2-\mu_2)]\,\hat{P}_1\,, \\ a_{22} &= (m_2-\mu_2)\,w_2\hat{P}_1 - (e_1+\varepsilon_1+e_2+\varepsilon_2)\,. \end{aligned} \tag{4.42}$$

The equilibrium with only species 1 is stable (i.e., species 2 cannot invade) if all the eigenvalues of the "community" matrix **A** have negative real parts (cf. Sect. 2.4). According to the Routh-Hurwiz criterion this requires *first* that the trace of the matrix is negative, i.e., $\mathrm{tr}(\mathbf{A}) = a_{11} + a_{22} < 0$, *and secondly* that the determinant of **A** is positive. These conditions are rather complicated mainly because of the large number of parameters, but in some simplified situations we can analyze the system. We will in the following give some results of Slatkin (1974).

First consider the simplest model in which the two species are symmetric with respect to their interaction, i.e., $m_1 = m_2 = m$; $\mu_1 = \mu_2 = \mu$; $e_1 = e_2 = e$; and $\varepsilon_1 = \varepsilon_2 = \varepsilon$. From the equality of the migration parameters using Eq. (4.38) we get that the trace is

$$\mathrm{tr}(\mathbf{A}) = -(m-\mu)\,(1-w_2)\,\hat{P}_1 - (2e+2\varepsilon)\,, \tag{4.43}$$

which is always negative. The determinant becomes

$$\det(\mathbf{A}) = -(m-\mu)\,\hat{P}_1 w_2[(m-\mu)\,\hat{P}_1 + 2e]\,, \tag{4.44}$$

which is negative whenever $m > \mu$ (state [1, 2] possible) and $w_2 > 0$ (state [1, 2] supply migrants). Thus in the completely symmetric case coexistence is always possible.

In the following we will, for convenience, assume that $w_2 = 1$. The "pure migration competition" model (Levins and Culver, 1971) assumes that $\varepsilon_1 = \varepsilon_2 = 0$; i.e., the only interaction between the species is that they hinder the migration of each other. In this model the condition that the trace is positive thereby allowing species 2 to invade is difficult to analyze. The other possibility allowing species 2 to invade, i.e., that the determinant is negative, is

$$R = (1 - e_2/m_2)/(1 - e_1/m_1) > \mu_2/m_2\,. \tag{4.45}$$

Condition [Eq. (4.45)] means that the relative decrease in the ability of species 2 to invade a patch already inhabited by species 1 should be less than the population size it can sustain when alone relative to species 1 [cf. Eq. (4.38)]. Thus, under migration competition, invasion is rather easy as long as $\mu$ is not too large and the two populations formed by the species when living alone are not too different. Reversing condition [Eq. (4.45)] we get a similar condition for the ability of species 1 to invade a system in which species 2 is established; we see that coexistence is possible for a broad range of parameter values. Note, however, that when $\mu$ approaches $m$ then we return to the invasion condition [Eq. (4.39)] for the model in which cohabitation is excluded. Thus for large values of $\mu$ the condition [Eq. (4.45)] becomes one that requires that the equilibrium sizes of the two monospecies systems should be nearly equal.

The last model we will consider is one which Levins and Culver (1971) called "extinction competition". Here migration is free to any patch (i.e., $\mu_i=0$) such that the interaction between the species is only due to the mutual exclusion from cohabited patches. Immigration of species 2 is possible if either the trace of **A** is positive, which is a strong condition on the parameters, *or* if the determinant of **A** is negative, which is the condition

$$[(1+m_1/m_2)R^{-1}-1]\,\varepsilon_2<\varepsilon_1+m_1+e_2\,, \tag{4.46}$$

where $R$ is the ratio on the left side of Eq. (4.45). The right side of Eq. (4.46) is always positive, such that the condition is fulfilled if

$$R=(1-e_2/m_2)/(1-e_1/m_1)>(1+m_1/m_2)\,, \tag{4.47}$$

and if this is not true, then condition [Eq. (4.46)] becomes

$$\varepsilon_2<[\varepsilon_1+m_1+e_2]/[(1+m_1/m_2)R^{-1}-1]\,. \tag{4.48}$$

Condition [Eq. (4.47) says that if the population size of species 2 when alone is much larger than that of species 1 when alone, then invasion is guaranteed. If Eq. (4.47) is fulfilled then the symmetric condition for the invasion of species 1 is not fulfilled and coexistence is not fulfilled. Condition [Eq. (4.48)] may be fulfilled for $\varepsilon_2$ sufficiently small. If we consider the symmetric condition for species 1 we see that the condition for coexistence is that the $\varepsilon$'s are sufficiently small.

We can now see that if condition [Eq. (4.47)] is fulfilled for the invasion of species 1 in a species 2 population (the symmetric condition of the one stated) then $R^{-1}$ is large and consequently condition Eq. (4.48)] requires that $\varepsilon_2$ be very small, so if either version of condition [Eq. (4.47)] is fulfilled, coexistence is unlikely.

The model [Eq. (4.37)] may be obtained from the general model [Eq. (4.40)] in two ways; by making $\mu_i=m_i$ or by letting $\varepsilon_i$ grow towards infinity. These possibilities correspond to the situation of competitive exclusion where either species introduced in a low number into a population of the other species will prohibit its increase. In this situation there cannot be coexistence except for the pathological case where the population sizes of each of the species are identical when alone in the system. Even in this case we would expect that a finite number of patches would lead to the random extinction of one or the other of the species.

Finally we give an example of a case that does not allow the persistent coexistence *within* a patch, but may allow regional coexistence. This is where one species excludes the other by competition; this could, e.g., correspond to $\mu_2=m_2$ and $\varepsilon_2>\varepsilon_1$, i.e., species 2 cannot invade a patch when species 1 has an established population, and if species 1 invades a patch inhabited by species 2, then species 1 has a competitive advantage. To take a simple example let $\varepsilon_1=\mu_1=0$, then species 1 can always invade a population of species 2, and species 2 can invade a population of species 1 if condition [Eq. (4.47)] is fulfilled, or if condition [Eq. (4.47)] is false then if

$$e_2+\varepsilon_2<e_1/[(1+m_1/m_2)R^{-1}-1] \tag{4.49}$$

where $R$ is the ratio on the left side of Eqs. (4.45) and (4.47). Thus in this system coexistence is possible in a situation where it would be impossible in a homogeneous population.

Coexistence of two species in a system of patches is most easily obtained when coexistence is possible for a limited time in each patch. Prolonged coexistence, however, is hindered by a decrease in the population sizes of both species due to competition. The model may therefore apply well to a group of competitors with a considerable niche overlap so that coexistence is impossible due to a vanishing stability of the coexistence equilibrium (see Sect. 3.4 and May, 1974). In a multispecies system we would not expect that all competitors should be found in any one patch but depending on the parameters of the model we could find a typical mean number of species in the patches. Thus an extension of these models may prove to apply well to the kind of biogeographical data which will be discussed in the following section.

Finally the evolution of interspecific antagonism should be mentioned in this connection. If, e.g., individuals of species 2 have adaptations which decrease $\mu_2$ or $\varepsilon_2$ they will increase their fitness within a cohabited patch. On the other hand, in patches colonized only by individuals of species 2 which possess adaptations increasing $\mu_1$ or $\varepsilon_1$, the species could be expected to sustain a population longer in this patch and thereby increase its fitness. These arguments, of course, rest on the assumption that individuals within a patch are more closely related than individuals on the average.

Thus antagonistic competition is not necessarily minimized like exploitative competion as discussed in Sect. 3.5 but may be favored by selection. Examples are the evolution of antibiosis in soil microorganisms, growth inhibitors in certain shrubs (Muller et al., 1964), and interspecific aggression in animals.

## 4.5 The Theory of Island Biogeography

The theory of island biogeography was originally developed to explain the species numbers of oceanic islands; the theory has since proven to have a much wider applicability for describing the faunas of isolates and for community ecology in general (Diamond, 1973; Simberloff, 1974; Lassen, 1975; Fenchel, 1975a; MacArthur and Wilson, 1963, 1967).

Basic to the theory is the empirical observation that species numbers of isolates show a positive correlation with area and a negative correlation with the distance from sources of colonization. This could in part be understood as an "under-saturation" of distant islands. However, observations on newly established isolates or isolates which have been defaunated artificially indicate a rapid recolonization which relatively quickly leads to a stable number of species.

The theory of island biogeography explains the species numbers of isolates as a dynamic equilibrium between the immigration of new species and the extinction of previously established species. It is intuitively clear that isolates will receive a higher number of immigrants if they are situated close to the source of immigrants or they are larger, thus constituting a larger target for immigrants. The probability of extinction of populations is believed to be larger in small isolates because of small total population sizes.

### 4.5.1 The Immigration and Extinction Curves

In order to discuss the theory of island biogeography a number of definitions must be made. The source of colonists is considered to have a total pool of $S^*$ species. By immigration to an isolate is meant the arrival of a *propagule* of a species which is not already represented on the isolate. By a propagule is meant a number of individuals sufficiently large to be potentially capable of establishing a population. Propagules of species which are already established on an isolate will, of course, also arrive at a certain rate. However, in accordance with MacArthur and Wilson (1967) we will use the term immigration only to describe the arrival of species to an isolate where the species is not already established.

The immigration rate, $I$, can be plotted against the number of species already present in an isolate. The simplest assumption is that $I$ is a linear function of the number of species present, $S$, so that $I$ takes a maximum value for $S=0$ and the value 0 for $S=S^*$. However, since it is likely that some species are better migrants or colonizers than others, we would expect $I$ to be relatively higher for low values of $S$ and relatively lower for high values of $S$ so that the immigration curve is concave (Fig. 60).

The extinction rate, $E$, is the rate of extinction of species which are established on the isolate. When $S=0$, then $E=0$. If the probability of extinction of the individual species populations is independent of $S$, $E$ would be a linearly increasing function of $S$. However, we would expect the probability of extinction of the species to increase with increasing $S$ since increasing competitive interactions will decrease the population sizes of each of the species. A concave extinction curve is therefore predicted (Fig. 60).

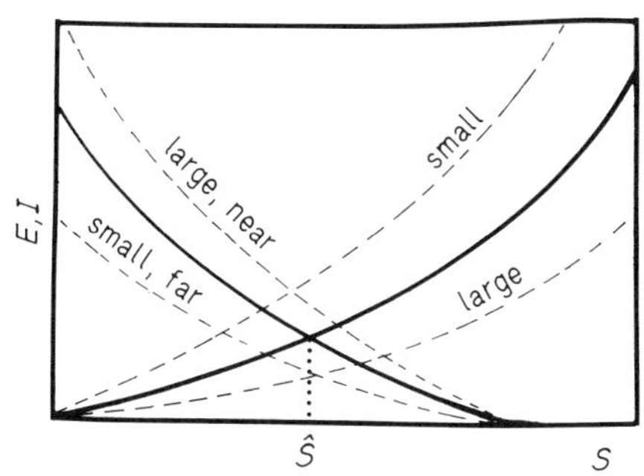


Fig. 60. The immigration and extinction as functions of species numbers for isolates of different sizes and distances from the source of colonization

The intersection of the extinction and the immigration curves determines the equilibrium number of species, $\hat{S}$, around which the actual number of species will vary. Thus, when $S=\hat{S}$, then $I=E$. If the actual number of species exceeds $\hat{S}$ the extinction rate will increase and $S$ will return to $\hat{S}$. Similarly, if $S$ falls below $\hat{S}$ the immigration rate will increase and the extinction rate will decrease.

We have not made any assumptions with respect to the qualitative composition of the fauna. It could be expected that different constellations of species on the isolate will yield different extinction and immigration curves. Any given isolate will, therefore, have a large number of alternative $I$ and $E$ curves corresponding to all the possible combinations of species drawn from the species pool $S^*$.

The actual extinction and immigration curves are difficult to observe in nature. Most evidence supporting the theory is based on a number of predictions deriving from the basic assumptions. These predictions include the area and the distance effects and their interaction, the constancy of species numbers but alternative species compositions of identical isolates (with respect to size and distance), the size of the variance of $\hat{S}$, and the shape of the immigration curve.

### 4.5.2 The Area and Distance Effects

The observed distance effect is one of the predictions of the theory; with increasing distance from the source of colonization the equilibrium number of species will decrease. The effect is especially well studied in the case of bird faunas on oceanic islands (MacArthur and Wilson, 1967; Diamond, 1973; see also Fig. 61).

Larger isolates have higher species numbers than small isolates. This also holds true when habitat samples of different sizes from a large, homogeneous area such as a mainland are compared. It has empirically been found that a good description of the species number on an island is given by the equation $\hat{S}=CA^z$, where $A$ is the area of the island and where $C$ and $z$ are constants. The parameter $C$ is the number of species found in a unit size area, $A=1$. The value of $C$ varies according to the particular habitat, the taxonomic position of the organisms in question and their population densities. The parameter $z$, which may be determined as the

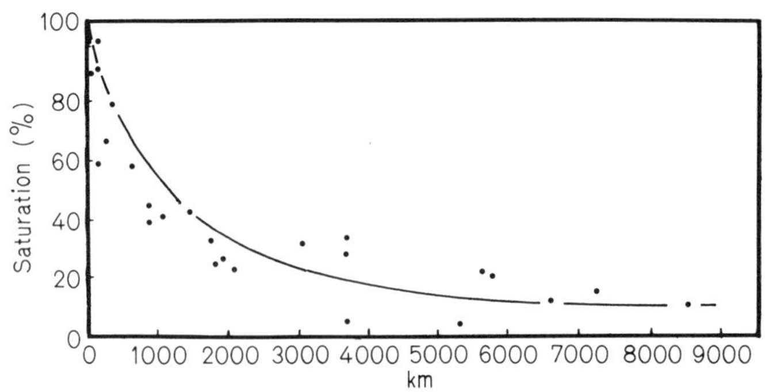


Fig. 61. The number of birds on islands (Moluccas, Melanesia, Oceania) as function of the distance from New Guinea given as percentage of the species numbers found on islands close to New Guinea. (After MacArthur and Wilson, 1963)

slope of the logarithmic regression line of $\hat{S}$ as a function of $A$, usually takes values between 0.2 and 0.35 (e.g., the number of species will double with an increase in area by a factor 10 for $z=0.301$). The factors influencing the value of $z$ will be discussed below, but let us first discuss the reasons for the area effect in general.

There are two main reasons for a higher equilibrium number of species on larger isolates. The probability of migration must increase as the target for immigration increases in size. The other reason is that the probability of extinction must increase with decreasing size of the isolate. As previously discussed (Sect. 4.3) all populations have a finite probability of going extinct. For density regulated populations this probability is high and approaches 1, but what is of interest here is that the expected time before extinction occurs decreases with decreasing population size. On small isolates the absolute population sizes will be smaller and, therefore, they will have a higher rate of extinction.

The size effect may be illustrated further by considering a large island as consisting of a number of more or less isolated habitat patches. On such an island, an extinct local subpopulation, comparable to a small island, will be relatively easily be replaced through migration from a surviving subpopulation somewhere else on the island. Total extinction on the large island therefore requires the simultaneous extinction of all the local subpopulations. A large isolate will therefore have a lower overall extinction rate, and consequently a higher value of $S$ than will a small isolate.

When a number of otherwise identical isolates which vary with respect to size and distance are compared, it is possible to distinguish between the effect of area and the effect of distance on the observed species numbers. Hamilton et al. (1964) found by multiple regression analysis that changes in area alone account for 80–90% of the variation in bird species numbers on some tropical archipelagos. Lassen (1975), in a study of freshwater snails in lakes, found no distance effect, i.e., area accounted for 100% of the variation when similar types of lakes are compared. This is explained by the high rate of dispersal in freshwater snails in conjunction with the high density of lakes in the studied region.

The equilibrium theory predicts that the relative increase in species when going from distant to close isolates is higher for small than for large isolates. It is also predicted that when going from small to large isolates the relative increase in the equilibrium species number will be higher for distant than for close isolates. In other words, higher extinction rates or lower immigration rates will yield higher values of $z$.

With the simplifying assumptions that $I$ and $E$ are linear functions of $S$ this effect is easily made intelligible by the following considerations.

Consider a large and a small isolate with the extinction curves

$$E=a_l S\,, \tag{4.50}$$

$$E=a_s S\,, \tag{4.51}$$

and with the immigration curve given by

$$I=b(S^*-S)\,, \tag{4.52}$$

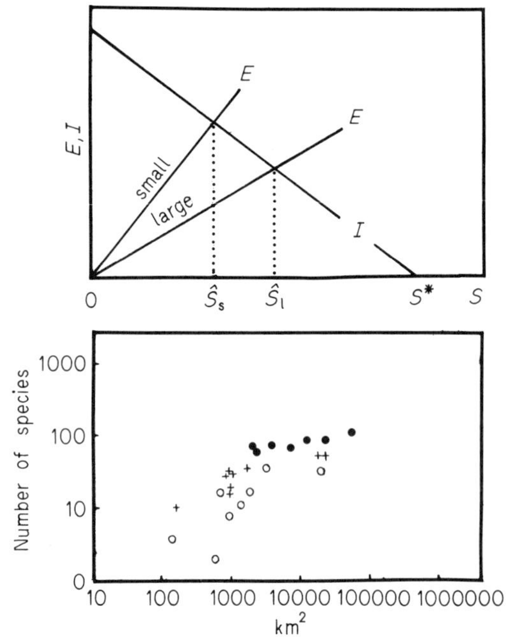


Fig. 62. *Above*: the increase in equilibrium species number when going from a small to a large island with the same distance from the mainland. *Below*: The species-area curve for the number of birds on three groups of Pacific islands lying at different distances from New Guinea; the most distant islands have the steepest curves. *Filled circles*: within 800 km; crosses: 800–3200 km; *open circles*: more than 3200 km. (After MacArthur and Wilson, 1967)

as shown on Figure 62. The two isolates will equilibrate at $\hat{S}_l$ and $\hat{S}_s$ species respectively. The relative increase in $S$ by going from the small to the large isolate is given by

$$(\hat{S}_l - \hat{S}_s)/\hat{S}_s\,. \tag{4.53}$$

By letting Eq. (4.52) equal Eqs. (4.50) and (4.51) respectively, we have

$$\begin{aligned} \hat{S}_l &= S^* b/(a_l + b)\,, \\ \hat{S}_s &= S^* b/(a_s + b)\,, \end{aligned} \tag{4.54}$$

which substituted into Eq. (4.53) yields

$$(\hat{S}_l - \hat{S}_s)/\hat{S}_s = (a_s - a_l)/(a_l + b)\,. \tag{4.55}$$

The total species pool, $S^*$, is the same for close and for distant isolates and the immigration curves will have lower slopes, i.e., lower values of $b$ for more distant isolates. From Eq. (4.55) it can be seen that the relative increase of $\hat{S}$ will be higher for lower values of $b$, i.e., for more distant islands. In a similar way it can easily be shown that the distance effect is greater for smaller than for larger islands.

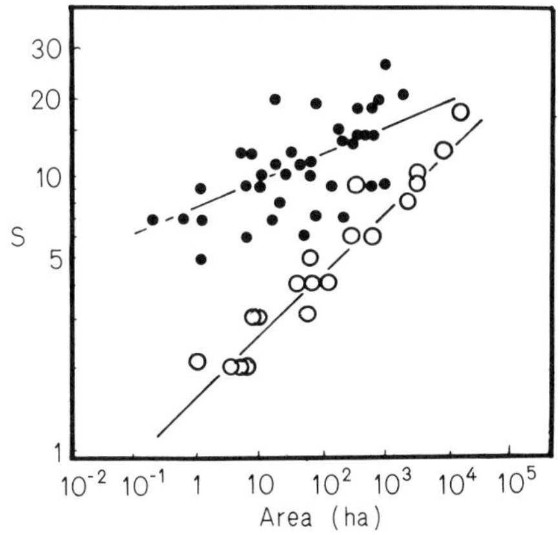


Fig. 63

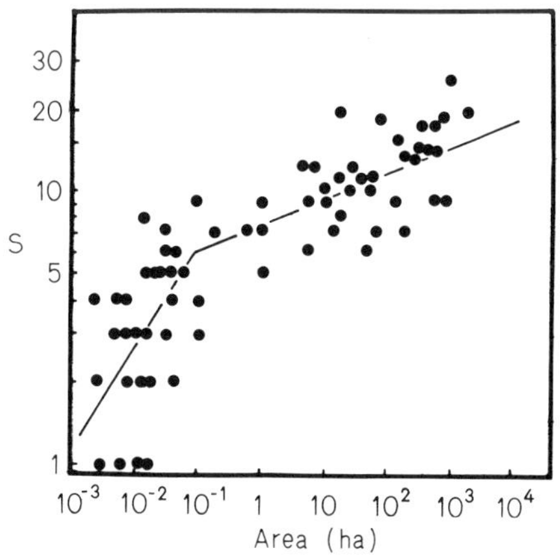


Fig. 64

Fig. 63. Species-area curves for freshwater snails in oligotrophic (*open circles*) and eutrophic (*filled circles*) ponds and lakes in Denmark. (After Lassen, 1975)

Fig. 64. Species-area curve for freshwater snails in eutrophic lakes, ponds and very small ponds. (After Lassen, 1975)

These predictions have been demonstrated an several occasions (an example is given on Figure 62). Lassen (1975) found a slope of the species-area curve of only 0.09 for freshwater snails in Danish eutrophic (nutrient rich) ponds and lakes varying in size from $10^{-1}$ to $10^4$ ha, but a slope of 0.25 for oligotrophic (nutrient poor) ponds and lakes of the same size range. The total species pool (i.e., the number of freshwater snails recorded from Denmark) is 34, which in only little more than found in the largest lakes of both categories (Fig. 63). The higher value of $z$ (but in general a lower value of $\hat{S}$) for oligotrophic lakes may be explained by a higher extinction rate due to the lower population densities relative to eutrophic lakes, such that an oligotrophic lake of a size comparable to a eutrophic lake actually constitutes a smaller "island." Note that the distribution of the snails is not directly governed by the physico-chemical properties characteristic of the two types of lakes since the largest lakes of both categories harbor about the same number of species which constitute the greater part of the total species pool.

The slopes of the species-area curves from mainland habitats are generally lower than those from isolates. This is in accordance with the considerations given above; immigration from one patch to another is generally facilitated relative to the immigration to, e.g., islands and the area effect thus should be smaller.

In the case of very small isolates, qualitative properties may result in values of $z$ differing from those found for larger isolates. Thus, Lassen (1975) found that the species-area slope for very small ponds ($<0.1$ ha) was very high, i.e., 0.37 (Fig. 64). This may be explained by the fact, that in very small ponds, several habitat types may be totally absent. A small pond may, for example, be entirely devoid of rooted vegetation or alternatively be totally covered by one kind of plant. When ponds are a little larger, more habitat types are added until essentially all types of freshwater habitats are represented within one pond. Above this size (which

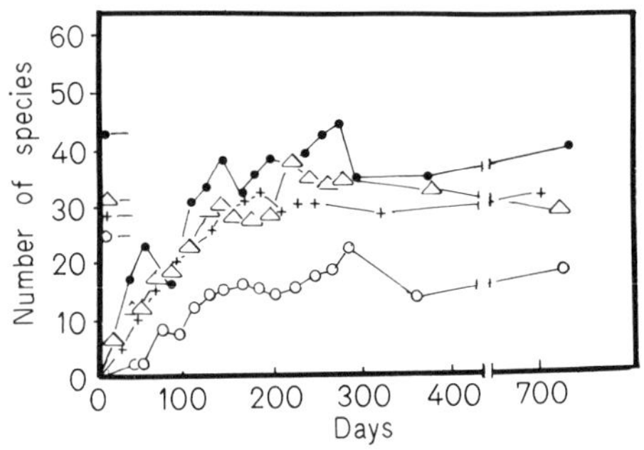


Fig. 65. The recolonization curves of 4 defaunated mangrove islands. The numbers of species prior to the defaunation are shown on the *y*-axis. (After Simberloff and Wilson, 1969)

apparently is around 0.1 ha) the value of $z$ takes a constant and lower value. The fact that very small ponds may freeze solid or dry up now and then, probably increases the extinction probability, and this effect will also lead to a higher value of $z$. MacArthur and Wilson (1967) give similar examples for very small islands devoid of forests.

Direct evidence from nature on the maintenance of an equilibrium number of species is less frequent. Diamond (1969) analyzed two censuses of the bird fauna on the Channel Islands off California, made at an interval of 50 years. While the total numbers of species on the individual islands had remained relatively constant, there had been a high number of replacements with other species; of the original species, 17 and 62% had disappeared on the largest and on the smallest islands respectively, to be replaced by an equal number of other species. From these figures a minimum estimate of the annual immigration and extinction rates at equilibrium could be made; these rates were from 0.3 to 1.2% of $\hat{S}$ for the largest and the smallest islands respectively.

Simberloff and Wilson (1969) defaunated a number of small mangrove islands in the Florida Keys and followed the recolonization curves. They found a rapid recolonization and the different islands returned to their original species numbers (mainly of microarthropods) within 3–4 months although the composition of each of the faunas differed more or less from the original one (Fig. 65). Lassen (pers. comm.) followed the colonization by snails in two small artificial ponds of known age and was able to follow the extinction of one species while it was being replaced by another species.

### 4.5.3 The Regulation of the Equilibrium Number of Species

The regression line of the species-area relationship represents the mean equilibrium values around which the actual values will be distributed. The theory of island biogeography predicts that during the initial, unsaturated stages following defaunation the variance/mean ratio will equal 1; as the equilibrium number of species is approached the ratio will decrease to values around 0.5 as a result of the regulation of the species number by two opposing forces. The formal argument is given by MacArthur and Wilson (1967). Thus variance to mean ratios smaller than unity for groups of identical isolates give support to the theory. Such data have been reported in the literature (e.g., Lassen, 1975).

### 4.5.4 The Colonization Curve

Under the simplifying assumption of linear $I$ and $E$ curves, the colonization curve can be constructed and the time taken for a given isolate to reach a certain fraction of the equilibrium number of species can be calculated.

Assume that $E = aS(t)$ and $I = b(S^* - S(t))$ where $S(t)$ is the number of species in the isolate at time $t$. The rate of change of $S$ is given by

$$dS/dt = I - E = bS^* - (a+b)\,S(t)\,. \tag{4.56}$$

By setting Eq. (4.56) equal to zero, we have

$$S^* = \hat{S}(a+b)/b\,,$$

which substituted into Eq. (4.56) yields

$$dS/dt = [\hat{S} - S(t)]\,(a+b)$$

or

$$d(\hat{S} - S(t))/dt = -[\hat{S} - S(t)]\,(a+b)\,,$$

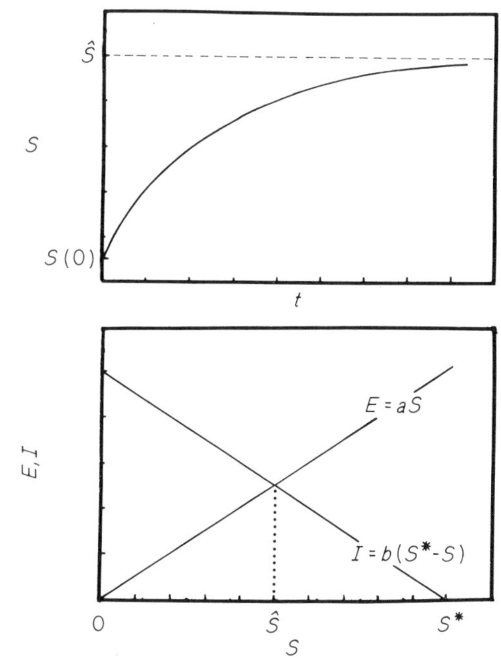


Fig. 66. The colonization curve calculated from Eq. (4.57) and the corresponding extinction and migration curves shown below

showing that the difference between $S(t)$ and the equilibrium species number decreases exponentially at the rate $(a+b)$. Therefore we have

$$S(t)=\hat{S}-[\hat{S}-S(0)]\exp[-t(a+b)]\,. \tag{4.57}$$

Equation (4.57) allows calculation of the colonization curve (Fig. 66) or the time taken to reach a certain percentage of the equilibrium species number for given immigration and extinction rates, or estimation of the latter from colonization curves. It should be remembered, however, that Eq. (4.57) is based on somewhat unrealistic assumptions. Attempts to estimate $I$ and $E$ on the basis of colonization curves were done by MacArthur and Wilson (1967) for data on the refaunation of the Krakatau Islands and by Lassen (1975) for colonization data on freshwater snails in artificial ponds.

### 4.5.5 The Significance of the Theory of Island Biogeography

The study of island faunas and floras has long been of importance in ecological, biogeographical and evolutionary research. The general ecological significance of the theory of island biogeography has recently been stressed by Diamond (1973) and by Simberloff (1974). In addition to being a predictive theory for the faunas of isolates, it has generated new insight into the nature of biotic communities by integrating concepts of competitive interactions, probabilities of extinction, and fugitive equilibria into a general theory of biotic communities.

# 5. The Concept of Biological Communities

The approach adopted in this book is to define parameters which describe how the biological and abiological environment affects populations. With simple assumptions about these parameters, we have then attempted to build predictive models which yield insight with respect to the distribution patterns of organisms in nature. This chapter will summarize the significance of the approach for understanding the concept of communities.

We have not taken any a priori attitude to the concept of biotic communities except for referring to the fact that within habitats, populations of different species may interact, and that other species form a component of the environment to which species adapt by natural selection. We have not considered any particular scale for the size of a community in terms of diversity or in terms of distribution in time or space; nor have we considered any criteria for the delimitation of communities. In this respect we are in accordance with Elton (1927) who states that "... the term animal communities is really a very elastic one, since we can use it to describe on one hand the fauna of equatorial forest, and on the other hand the fauna of a mouse's caecum" (see also Fig. 67). We agree with this elastic definition although, as discussed later, the controversy in the literature about the "existence" of communities in part stems from a question of scale.

To Odum (1971) "the community concept is one of the most important principles in ecological thought and ecological practice". Somewhat less modest are Allee et al. (1949) who write: "The community concept of modern ecology is one of the most fruitful ideas contributed by biological science to modern civilization." Considering the title of our book it seems appropriate, on this background and at this stage, to comment on the concept of communities and on how the book relates to it.

The concept of biological communities stems primarily from the fundamental observation that animals and plants are not homogeneously distributed on the globe or even within restricted geographical areas, and that the occurence of different species is often correlated. Thus, within a given geographical region the designations, e.g., beech forest or coral reef predict the presence of many other species within the two types of communities in addition to the organisms which have given their names to the associations. In part, the discrete distribution of assemblages of organisms sometimes found is trivial. Certain qualities of the physical and the chemical environment are often found within sharply defined areas, and giraffes and elephants would not be expected in the sea just as whales and fish are rare on the savannah.

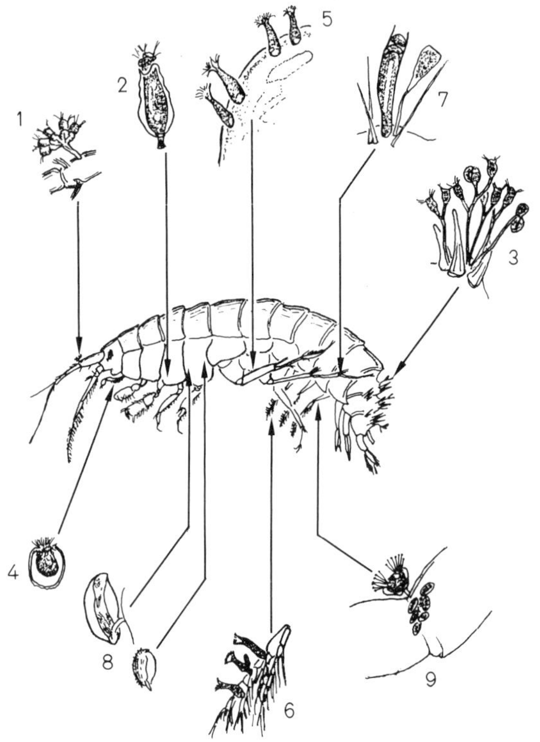


Fig. 67. The community of ciliate protozoans associated with the marine amphipod *Gammarus locusta*. *1–4*: Four species of very site-specific peritrich ciliates; no 3 is parasitized by another ciliate, *Hypocoma parasitica*. *5–6*: Two species of chonotrich ciliates living on the gills and the abdominal feet respectively. The peritrichs as well as the chonotrichs are filterfeeders, utilizing the water currents created by the host. *7*: A parasitic ciliate living in some way on the skin glands of the host. *8*: Two free-swimming ciliates presumably living on blugreen algae attached to the host. *9*: A suctorian catching other ciliates, and the cysts of a ciliate (*Gymnodinioides*) which excysts at moulting of the host to feed in the moult. (After Fenchel, 1965)

The various theories of biological communities developed largely from studies on particular ecological associations. The theories are often flavored by special properties of these systems, as well as by the general philosophical attitude of the different ecologists. Thus, communities have at one extreme been viewed as a kind of "supra organism" with a characteristic structure and development, with homeostatic properties and as always being sharply delimited from neighboring communities. Others have described communities as arbitrarily defined associations of organisms which are often found together due to their preferences for certain qualities of the physical environment. The first mentioned "holistic" view, stressing the biological interactions between species and communities as "natural units", was first proposed by Möbius (1877) who coined the word "biocoenosis" in a study of oyster beds of the North Sea.

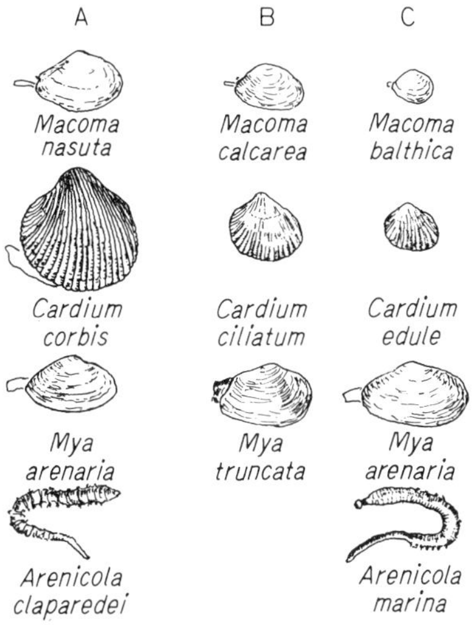


Fig. 68 A–C. The dominant invertebrates in shallow sand bottoms of the Pacific Northwest of USA (A), Greenland waters (B) and Danish waters (C). The same genera (albeit mainly different species) and thus the same ecological types are found, demonstrating Thorson's concept of "parallel animal communities". (Modified after Thorson, 1957)

However, the whole range of viewpoints is usually represented among students of each of the different kinds of associations. [To be sure, some ecologists who have their main experience with desert habitats (e.g., Bodenheimer, 1958) where it is difficult to conceive that the scattered flora can be a part of a system structured by biological interactions, or with pests (Andrewartha and Birch, 1954), have not found any value in the community concept whatsoever.] Within the study of "plant sociology", Braun-Blanquet (1932, 1951) and Clements (1905, 1916) have supported the idea of the community as a fundamental biological unit whereas, e.g., Gleason (1926) and Whittaker (1951) disregarded communities as discrete units.

A quite similar controversy is found in the literature on marine level bottom communities. After the turn of the century, Petersen (1913, 1918) divided the faunal associations on the sea bottom of Danish waters into "benthic communities" on a purely empirical basis. The classification and nomenclature of Petersen are still widely used by marine biologists and they are obviously useful for characterizing faunal assemblages, not only in Northern European waters but also in other parts of the world (see Fig. 68). Following Petersen, many ecologists took up the study of marine communities. Of those who studied complex and patchy habitats of fjords, estuaries, and the littoral, some gave a formal description of a large number of "communities" (e.g., Gislén, 1930). Other authors (e.g., Linroth, 1935; MacGinitie, 1939) took the consequence of the confusing picture emerging from their studies, and disregarded the concept of benthic communities in spite of the fact that relatively homogeneous and fairly well defined faunal assemblages are actually found in the sediments of more open waters. Some scientists (notably Thorson, 1957; see also references therein) stressed the benthic communities as "biological units" or "biocoenoses"; others advocated the physical factors (salinity, mechanical properties of the sediment, temperature) as governing the distribution of species and consequently stress the absence of discontinuities between communities.

The two latter mentioned points of view imply that given sufficient data on environmental factors, then the faunal composition can always be predicted. This implication has led to various suggestions as to how to make formal rules for delimiting communities and—not least in the age of the computer—to multifactor correlations and the development of similarity indexes based on quantitative samples with large numbers of species (e.g., Fager, 1963; Lie and Kelley, 1970). Without yielding much biological information this work tends to weaken the practical value of Petersen's benthic communities, which are best applied with common sense, tact and biological intuition.

The problem concerning the nature of biological communities as formulated by, e.g., the schools of plant sociology or of marine communities is resolved or rather rendered a nonproblem when the results of theoretical ecology are considered (and the results of natural history as well; many "community ecologists" have not considered that very simple experiments and observations could be as fruitful for studying the importance of biological interactions as could complex statistical analyses of quantitative data).

First of all it is clear that the biological *as well as* the abiological environment is important for explaining the distribution of organisms. There exist large amounts of evidence showing that more or less specific predators are important among, e.g., benthic animals, and a long list of mutualistic and commensalistic relationships are known. Also character divergence and habitat selection, which indicate coadaptation, and phenomena like ecological compression and expansion show that we could expect—as we do find—that distribution patterns of many species are positively or negatively correlated, conditioned on the physical environment.

On the other hand, the results derived in Chapters 2, 3 and 4 suggest that the number of species with strong biological interactions is limited within any one habitat. Thus, it may be conjectured that any community consists of small groups of species with relative strong interactions (such as prey-predator interactions, competition, mutualism), which coexist with a larger number of species with which there are only weak or no interactions. We have seen that prey–predator systems are endowed with some inherent stability but that the number of possible trophic levels is restricted. This limit is set by the resources available to the system in conjunction with demographic and environmental stochasticity which sets a limit to how small populations can be before the probability of extinction becomes very high. Within trophic levels strong competitive interactions are inherently unstable and the number of species with strong competitive interactions must be relatively small even if we allow for several niche dimensions. Finally, we have seen that a system of competitors which is stable will also be stable if some of the competing species are absent. It is therefore to be expected that the majority of species within an area have only weak or no interactions. This accounts for the fact that the occurrences of species are often not correlated along environmental gradients and that sharp boundaries between communities are frequently not found.

Finally, when phenomena related to geographical subdivision of populations such as migration-extinction equilibria are considered, it is clear that the composition of a community will never be totally predictable on the basis of environmental factors and this will be increasingly so in complex and patchy habitats (archipelagos in a wide sense: lakes, estuaries, topographically complex landscapes) and in environments characterized by unpredictable resources.

Thus the question of whether communities are "real, discrete biological units" is in fact not a real problem when we consider communities as the function of the component species populations and the environment. It is popular to consider ecological systems as a special "organizational level" (analogous to the cellular level, the level of the individual and the level of species populations) for which theories and predictive models can be made independently of the behavior of the individual components. In studies of energy flow and element cycling this has proved a useful approach and the practical value of studying, e.g., the effect of adding phosphate to a lake on the total photosynthetic activity is indisputable. However, the concept of organizational levels with their own properties and a special theory for such levels not based on the behavior of single species populations may be misleading and result in false analogies. An example has been the application of information theory to communities, leading to peculiar and somewhat obscure predictions like: "... that the community seeks to gain and does gain information from the environment, only to use such accumulated information to block any further assimilation of information" (Margalef, 1968). This sort of theory may at the best yield empirically correct descriptions (formulated in teleological expressions; communities certainly do not "seek" anything whatsoever); it may also generate predictions which are persistent in the literature although they cannot be sustained as empirically true generalities [cf. the discussion by May (1973a) on stability and diversity].

The natural units of ecological systems are, after all, the individual species. A community changes by the exclusion or invasion of species just like the single species population changes by the birth or the death of individuals. However, the single species populations, as Mendelian populations, share pools of genetic material which determine the appearance of the species in the future (Dobzhansky, 1970). Therefore, the removal or addition of individuals because of their genotype, i.e., natural selection, changes the genetic composition and thereby the future appearance of the *individuals* in the population. This unifying principle of Mendelian populations allows for a predictive theory of the composition and the evolution of a species at the population level. Such a unifying principle does *not* exist for ecological systems which consequently must be described in terms of their components. Nevertheless, it is often valuable to describe generalizations about the behavior of whole communities, e.g., patterns of successions and diversity, but any valid explanation of such generalizations must be based on the behavior of the single species populations and their interactions.

Elton's (1927) pioneering book, based on so much ecological knowledge and intuition, contains (in addition to a good deal more) a substantial part of the ideas and insight discussed in this book, including the concept of the niche and the importance of trophic relationships and of migrations for determining the structure of biological communities as we see them in nature. It may, therefore, well be asked what are the most important advances in the development of ideas in ecological theory since the publication of Elton's book nearly half a century ago. Several new ideas have, after all, emerged; this should be clear from the chapters of the present book. One aspect which we will stress here is the mathematical formulation of the theory. This is important for two reasons. A mathematical formulation of a theory (made in conjunction with biological intuition) is more resistant against false conclusions and makes it clearer whether a problem is understood or not. Secondly, a mathematical formulation leads to quantitative (or in ecology, rather order of magnitude) predictions which can be tested in nature; the theory of limiting similarity is a good example, where verbal theory alone would not have yielded the insight of the mathematically formulated theory.

Many biologists, especially among those who have been trained to distinguish details and to study the diversity of life (and this is after all a necessary component of the training of a good biologist since one of our most important goals is to explain the diversity of life), discard an approach like the one adopted in this book as an oversimplification. We acknowledge (and enjoy) the immense complexity of nature; still we feel that the theory represented here constitutes an, albeit tiny, element of a true understanding of the real world beyond a purely descriptive approach.

# References


Allee, W.C., Emerson, A.E., Park, O., Park, T., Schmidt, K.P.: Principles of Animal Ecology. Philadelphia: Saunders Co., 1949

Anderson, W.W.: Genetic equilibrium and population growth under density-regulated selection. Am. Nat. **105**, 489–498 (1971)

Andrewartha, H.G., Birch, L.C.: The Distribution and Abundance of Animals. Chicago: Univ. Chicago Press, 1954

Ayala, F.J.: Experimental invalidation of the principle of competitive exclusion. Nature **224**, 1076–1079 (1969)

Barker, J.S.F.: Natural selection for coexistence or competitive ability in laboratory populations of *Drosophila*. Egypt. J. Genet. Cyt. **2**, 288–315 (1973)

Bick, H.: Die Sukzession der Organismen bei der Selbstreinigung von organisch verunreinigtem Wasser unter verschiedenen Milieubedingungen. Min. ELF Nordrhein/Westfalen, Düsseldorf, 1964

Bodenheimer, F.S.: Animal Ecology Today. Den Haag: W. Junk., 1958

Braun-Blanquet, J.: Plant Sociology: the Study of Plant Communities. New York: McGraw-Hill, 1932

Braun-Blanquet, J.: Pflanzensoziologie. Wien: Springer, 1951

Brown, J., Wilson, E.O.: Character displacement. Syst. Zool. **5**, 49–64 (1956)

Cameron, W.A.: Mammals of the islands in the Gulf of St. Lawrence. Nat. Museum, Canada, Bull. **154**, 1–165 (1958)

Charlesworth, B.: Selection in density-regulated populations. Ecology **52**, 469–474 (1971)

Clarke, B.: Density dependent selection. Amer. Nat. **106**, 1–13 (1972)

Clements, F.E.: Research Methods in Ecology. Lincoln, Nebraska: Univ., 1905

Clements, F.E.: Plant succession: analysis of the development of vegetation. Publ. Carnegie Inst. **242**, 1–512 (1916)

Cody, M.L.: A general theory of clutch size. Evolution, **20**, 174–184 (1966)

Cohen, D.: Optimizing reproduction in a randomly varying environment. J. Theoret. Biol. **16**, 1–14 (1967)

Cohen, J.E.: A Markov contingency-table model for replicated Lotka-Volterra systems near equilibrium. Am. Nat. **104**, 547–560 (1970)

Cole, L.C.: The population consequences of life history phenomena. Quart. Rev. Biol. **29**, 103–137 (1954)

Corliss, J.O.: History taxonomy, ecology, and evolution of species of *Tetrahymena*. In: Biology of *Tetrahymena*. Elliott, A.M. (ed.): Stroudsburg, Pennsylvania: Dowden, Hutchinson and Ross, 1973, pp. 1–55

Crombie, A.C.: Further experiments on insect competition. Proc. Roy. Soc. London, **133 B**, 76–109 (1946)

Crow, J.F., Kimura, M.: An Introduction to Population Genetics Theory. New York: Harper and Row 1970

Diamond, J.M.: Avifaunal equilibria and species turnover rates on the Channel Islands of California. Nat. Acad. Sci. **64**, 57–63 (1969)

Diamond, J.M.: Distributional ecology of New Guinea birds. Science **179**, 759–769 (1973)

Dobzhansky, T.: Genetics of the Evolutionary Process. New York: Columbia Univ. Press, 1970

Elton, C.: Animal Ecology. London: Sidgwick and Jackons, 1927

Elton, C.: The Ecology of Animals. London: Methuen, 1933
Ewens, W. J.: Population Genetics. London: Methuen, 1969
Fager, E. W.: Communities and organisms. In: The Sea. Hill, M. N. (ed.): New York: Wiley Interscience, 1963, pp. 415–437
Feldman, M. W., Roughgarden, J.: A populations stationary distribution and chance of extinction in a stochastic environment with remarks on the theory of species packing. Theoret. Popul. Biol. **7**, 197–207 (1975)
Fenchel, T.: On the ciliate fauna associated with the marine species of the amphipod genus *Gammarus*. Ophelia **2**, 281–303 (1965)
Fenchel, T.: The ecology of marine microbenthos II. The food of marine benthic ciliates. Ophelia **5**, 73–121 (1968)
Fenchel, T.: Intrinsic rate of natural increase: the relationship with body size. Oecologia (Berl.) **14**, 317–326 (1974)
Fenchel, T.: Factors determining the distribution patterns of mud snails (Hydrobiidae). Oecologia (Berl.) **20**, 1–17 (1975a)
Fenchel, T.: Character displacement and coexistence in mud snails (Hydrobiidae). Oecologia (Berl.) **20**, 19–32 (1975b)
Fenchel, T., Kofoed, L. H.: Evidence for exploitative interspecific competition in mud snails (Hydrobiidae). Oikos **27**, 367–376 (1976)
Fisher-Piette, E.: Sur l'équilibre de faunas: interactions des moules, des poupres et des cirripedes. C. R. Soc. Biogeogr. **92**, 47–48 (1934)
Gadgil, M., Solbrig, O. T.: The concept of $r$- and $K$-selection: evidence from wild flowers and some theoretical considerations. Am. Nat. **106**, 14–31 (1972)
Gantmacher, F. R.: The Theory of Matrices. New York: Chelsea Publ. Co., 1960, Vol. I.
Gantmacher, F. R.: The Theory of Matrices. New York: Chelsea Publ. Co., 1971, Vol. II.
Gause, G. F.: The Struggle for Existence. New York: Hafner Publ. Co. 1934 (Reprinted in 1964)
Gause, G. F.: Vérifications expérimentales de la théorie mathématique de la lutte pour la vie. Hermann et C^ie^, Paris: Actualités scientifiques et industrielles 277, 1935
Gilpin, M. E.: A model of the predator-prey relationship. Theoret. Popul. Biol. **5**, 333–334 (1974)
Gislén, T.: Epibioses of the Gullmar Fjord. I–II. In: Kristinebergs Zoologiska Station 1877–1927, **3**, 1–123 and **4**, 1–380 (1930)
Gleason, H. H.: The individualistic concept of the plant association. Boll. Torrey Bot. Club. **53**, 7–26 (1926)
Göbber, F., Seelig, F. F.: Conditions for the application of the steady-state approximation to systems of differential equations. J. Math. Biol. **2**, 79–86 (1975)
Grinnell, J.: The niche-relationship of the California trasher. Auk. **34**, 427–433 (1917a)
Grinnell, J.: Fields tests of theories concerning distributional control. Am. Nat. **51**, 115–128 (1917b)
Hairston, N. G., Tinkle, D. W., Wilbur, H. M.: Natural selection and parameters of population growth. J. Wildlife Manage. **34**, 681–690 (1970)
Hamilton, T. H., Barth, R. H., Jr., Rubinoff, I.: The environmental control of insular variation in bird species abundance. Proc. Nat. Acad. Sci. U.S. **52**, 132–140 (1964)
Harris, T.: The Theory of Branching Processes. Berlin: Springer-Verlag, 1963
Hespenheide, H. A.: Ecological inferences from morphological data. Ann. Rev. Ecol. Syst. **4**, 213–229 (1973)
Holling, C. S.: The functional response of predators to prey density and its role in mimicry and population regulation. Mem. Ent. Soc. Can. **45**, 1–60 (1965)
Horn, H. S., MacArthur, R. H.: Competition among fugitive species in a harlequin environment. Ecology **53**, 749–752 (1972)
Huffaker, C. B.: Experimental studies on predation: dispersion factors and predator prey oscillations. Hilgardia **27**, 343–383 (1958)
Hutchinson, G. E.: Circular causal systems in ecology. Ann. N. Y. Acad. Sci. **50**, 221–246 (1948)
Hutchinson, G. E.: Copepodology for the ornithologist. Ecology **32**, 571–577 (1951)
Hutchinson, G. E.: Concluding remarks. Cold Spring Harbor Symposia on Quantitative Biology **22**, 415–427 (1957)

Hutchinson, G. E.: Homage to Santa Rosalia or why are there so many kinds of animals? Am. Nat. **93**, 145–159 (1959)
Janzen, D. H.: Herbivores and the number of tree species in tropical forests. Am. Nat. **104**, 501–528 (1970)
Jones, N. S.: Marine bottom communities. Biol. Rev. **25**, 283–313 (1950)
Karlin, S.: A First Course in Stochastic Processes. New York: Academic Press 1966
Karlin, S., McGregor, I.: Polymorphisms for genetic and ecological systems with weak coupling. Theoret. Pop. Biol. **3**, 210–238 (1972)
Keiding, N.: Extinction and growth in random environments. Theoret. Popul. Biol. **8**, 49–63 (1975)
Kohn, A. J.: Microhabitats, abundance and food of *Conus* on atoll reefs in the Maldive and Chagos islands. Ecology **49**, 1046–1062 (1968)
Kolmogoroff, A.: Sulla teoria di Volterra della lotta per l'esistenza. Giornale dell' Istituto Italiano degli Attuari **7**, 74–80 (1936)
Kormondy, E. J.: Concepts of Ecology. Englewood Cliffs, New Jersey: Prentice-Hall, 1969
Lack, D.: The significance of clutch size. Ibis **87**, 302–352 (1947)
Lassen, H. H.: The diversity of freshwater snails in view of the equilibrium theory of island biogeography. Oecologia (Berl.) **19**, 1–8 (1975)
Lawton, J. H., Beddington, J. R., Bonser, R.: Switching in invertebrate predators. In: Ecological Stability. Usher, M. B., Williamson, M. H. (eds.). London: Chapman and Hall, 1974, pp. 141–158
Leslie, P. H.: Some further notes on the use of matrices in population mathematics. Biometrika **35**, 213–245 (1948)
Levin, S. A.: Dispersion and population interactions. Am. Nat. **108**, 207–228 (1974)
Levin, S., Paine, R. T.: Disturbance, patch formation, and community structure. Proc. Nat. Acad. Sci. **71**, 2744–2747 (1974)
Levins, R.: Toward an evolutionary theory of the niche. In: Evolution and Environment. Drake, E. T. (ed.). New Haven, Conn: Yale Univ. 1968, pp. 325–340
Levins, R., Culver, D.: Regional coexistence of species and competition between rare species. Proc. Nat. Acad. Sci. U.S. **68**, 1246–1248 (1971)
Lewontin, R. C.: Selection for colonizing ability. In: The Genetics of Colonizing Species. Baker, H. G., Stebbins, G. L. (eds.). New York: Academic Press, 1965, pp. 77–94
Lewontin, R. C.: The Genetic Basis of Evolutionary Change. New York: Columbia Univ. Press, 1974
Lewontin, R. C., Cohen, D.: On population growth in a randomly varying environment. Proc. Nat. Acad. Sci. U.S. **62**, 1059–1060 (1969)
Lie, U., Kelley, J. C.: Benthic infauna communities off the coast of Washington and in Puget Sound: identification and distribution of the communities. J. Fish. Res. Bd. Canada **27**, 621–651 (1970)
Linroth, A.: Die Associationen der Marinen Weichböden. Zool. Bidrag Uppsala **15**, 331–366 (1935)
Lotka, A. J.: Elements of Physical Biology. Baltimore: Williams and Wilkins, 1925, 460 pp. (Reprinted 1956 as: Elements of Mathematical Biology. Dover Publications).
MacArthur, R. H.: Some generalized theorems of natural selection. Proc. Nat. Acad. Sci. U.S. **48**, 1893–1897 (1962)
MacArthur, R. H.: Species packing and competitive equilibrium for many species. Theoret. Popul. Biol. **1**, 1–11 (1970)
MacArthur, R. H.: Geographical Ecology. New York: Harper and Row, 1972
MacArthur, R. H., Levins, R.: Competition, habitat selection and character displacement in a patchy environment. Proc. Nat. Acad. Sci. U.S. **51**, 1207–1210 (1964)
MacArthur, R. H., Levins, R.: The limiting similarity, convergence and divergence of coexisting species. Am. Nat. **101**, 377–385 (1967)
MacArthur, R. H., MacArthur, J.: On bird species diversity. Ecology **42**, 594–598 (1961)
MacArthur, R. H., Wilson, E. O.: An equilibrium theory of insular zoogeography. Evolution **17**, 373–387 (1963)
MacArthur, R. H., Wilson, E. O.: The Theory of Island Biogeography. Princeton: Princeton Univ. Press, 1967

MacGinitie, G.E.: Littoral marine communities. Am. Midl. Nat. **21**, 28–55 (1939)
Malthus, T.B.: First Essay on Population. 1798. (Reprinted 1926: London: Macmillan).
Margalef, R.: Perspectives in Ecological Theory. Chicago: Univ. Chicago Press 1968
May, R.M.: Limit cycles in predator-prey communities. Science **177**, 900–902 (1972)
May, R.M.: Stability and Complexity in Model Ecosystems. Princeton: Princeton Univ. Press, 1973a
May, R.M.: Stability in randomly fluctuating versus deterministic environments. Am. Nat. **107**, 621–650 (1973b)
May, R.M.: On the theory of niche overlap. Theoret. Popul. Biol. **5**, 297–332 (1974)
May, R.M.: Biological populations obeying difference equations: stable points, stable cycles and chaos. Theoret. Biol. **51**, 511–524 (1975)
Maynard Smith, J.: Models in Ecology. Cambridge: Cambridge Univ. Press, 1974
Maynard Smith, J., Slatkin, M.: The stability of predator prey systems. Ecology **54**, 384–391 (1973)
Mertz, D.B.: The mathematical demography of the California condor population. Am. Nat. **105**, 437–453 (1971)
Möbius, K.: Die Auster und die Austern-Wirtschaft. Wiegandt, Hempl und Parey, Berlin. 1877. English Transl.: Rept. U.S. Fish Comm., 1880. pp. 683–751
Monod, J.: La technique de culture continue, théorie et applications. Ann. Inst. Pasteur **79**, 390–410 (1950)
Moore, J.A.: Competition between *Drosophila melanogaster* and *Drosophila simulans*. I. Population cage experiments. Evolution **6**, 407–420 (1952)
Mountford, M.D.: The significance of clutch size. In: The Mathematical Theory of the Dynamics of Biological Populations. Bartlett, M.S., Hiorns, R.W. (eds.). London: Academic Press, 1973
Muller, C.H., Muller, W.H., Haines, B.L.: Volatile growth inhibitors produced by aromatic shrubs. Science **143**, 471–473
Murdoch, W.W.: Switching in general predators: experiments on predator specificity and stability of prey populations. Ecol. Monogr. **39**, 335–354 (1969)
Muus, B.J.: The fauna of Danish estuaries and lagoons. Medd. Danm. Fisk. Havunders. **5** (1), 1–316 (1967)
Nicholson, A.J.: An outline of the dynamics of animal populations. Aust. J. Zool. **2**, 9–65 (1954)
Nicholson, A.J., Bailey, V.A.: The balance of animal populations. Part I. Proc. Zool. Soc. London **3**, 551–598 (1935)
Nicolaisen, W., Kanneworff, E.: On the burrowing and feeding habits of the amphipods *Bathyporeia pilosa* Lindström and *Bathyporeia sarsi* Watkin. Ophelia **6**, 231–250 (1969)
Novick, A., Szilard, L.: Experiments with the chemostat on spontaneous mutations of bacteria. Proc. Nat. Acad. Sci. U.S. **36**, 708–719 (1950)
Ockelmann, K.W.: Development types in marine bivalves and their distribution along the Atlantic coast of Europe, pp. 25–35. In: Proc. First Europ. Malac. Congr. (1962), Cox, L.R., Peake, J.S. (eds.). London: Conchological Society of Great Britain and Ireland and Malacological Society, London (1965)
Odum, E.P.: Fundamentals of Ecology. Philadelphia: Saunders, 1971
Paine, R.T.: Food web complexity and species diversity. Am. Nat. **100**, 65–76 (1966)
Park, T.: Beetles, competition and populations. Science **138**, 1369–1375 (1962)
Pearl, R., Reed, L.S.: On the rate of growth of the population of the United States since 1790 and its mathematical representation. Proc. Nat. Acad. Sci. U.S. **6**, 275–288 (1920)
Petersen, C.G.J.: Valuation of the sea. II. The animal communities of the sea bottom and their importance for marine zoogeography. Rep. Danish Biol. Stat. **21**, 1–44 (1963)
Petersen, C.G.J.: The sea bottom and its production of fishfood. A survey of the work done in connection with the valuation of the Danish waters from 1883–1917. Rep. Danish Biol. Stat. **25**, 1–62 (1918)
Pianka, E.R.: Latitudinal gradients in species diversity: a review of concepts. Am. Nat. **100**, 33–46 (1966)
Pianka, E.R.: Sympatry of desert lizards (*Ctenotus*) in Western Australia. Ecology **50**, 1012–1030 (1969)

Pianka, E.R.: *r* and *K* selection or *b* and *d* selection? Am. Nat. **106**, 581–588 (1972)
Pianka, E.R.: The structure of lizard communities. Ann. Rev. Ecol. Syst. **4**, 53–74 (1973)
Pianka, E.R.: Evolutionary Ecology. New York: Harper and Row, 1974
Pielou, E.C.: An Introduction to Mathematical Ecology. New York: Wiley and Sons, 1969
Pinkster, S., Dennert, A.L., Stock, B., Stock, J.H.: The problem of European freshwater populations of *Gammarus duebeni* Liljeborg, 1852. Bijdr. Dierk. **40**, 116–147 (1970)
Poulsen, E.T.: A mathematical model for population regulation. I. Preprint Series 1974/1975 No 19. Matematisk Institut, Aarhus Universitet, 1975
Prout, T.: The estimation of fitnesses from genotypic frequencies. Evolution **19**, 546–551 (1965)
Remane, A., Schlieper, C.: Biology of Brackish Water. New York: Wiley and Sons, 1971
Rescigno, A., Richardson, I.W.: The struggle for life: I. Two species. Bull. Math. Biophysics. **29**, 377–388 (1967)
Robertson, F.W., Reeve, E.: Studies in quantitative inheritance. I. The effects of selection of wing and thorax length in *Drosophila melanogaster*. J. Genet. **50**, 414–448 (1952)
Rosenzweig, M.L., MacArthur, R.: Graphical representation and stability conditions of predator-prey interactions. Am. Nat. **97**, 209–223 (1963)
Roughgarden, J.: Density-dependent natural selection. Ecology **52**, 453–468 (1971)
Roughgarden, J.: Evolution of niche width. Am. Nat. **106**, 683–718 (1972)
Roughgarden, J.: Niche width: biogeographic patterns among *Anolis* lizard populations. Am. Nat. **108**, 429–442 (1974a)
Roughgarden, J.: Species packing and the competition function with illustration from coral reef fish. Theoret. Popul. Biol. **5**, 163–186 (1974b)
Roughgarden, J.: Species packing and faunal buildup: an evolutionary approach based on density dependent natural selection. Manuscript, 1975
Roughgarden, J., Feldman, M.: Species packing and predation pressure. Ecology **56**, 489–492 (1975)
Sanders, H.L.: Marine benthic diversity: a comparative study. Am. Nat. **102**, 243–282 (1968)
Schoener, T.W.: The evolution of bill size differences among sympatric congeneric species of birds. Evolution **19**, 189–213 (1965)
Schoener, T.W.: The *Anolis* lizards of Bimini: resource partitioning in a complex fauna. Ecology **49**, 704–726 (1968)
Simberloff, D.S.: Equilibrium theory of island biogeography and ecology. Ann. Rev. Ecol. Syst. **5**, 161–182 (1974)
Simberloff, D.S., Wilson, E.O.: Experimental zoogeography of islands: the colonization of empty islands. Ecology **50**, 278–296 (1969)
Skellam, J.G.: Random dispersal in theoretical populations. Biometrika **38**, 196–218 (1951)
Slatkin, M.: Competition and regional coexistence. Ecology **55**, 128–134 (1974)
Slobodkin, L.B.: Growth and Regulation of Animal Populations. New York: Holt, Rinehart and Winston, 1961
Slobodkin, L.B.: Ecological populations of hydrida. J. Anim. Ecol. **33** (Suppl.) 131–148 (1964)
Smith, C.C., Fretwell, S.D.: The optimal balance between size and offspring. Am. Nat. **108**, 499–506 (1974)
Stern, K., Roche, L.: Genetics of Forest Ecosystems. Berlin: Springer 1974
Strandgaard, H.: The roe deer (*Capreolus capreolus*) population at Kalø and the factors regulating its size. Danish Rev. Game Biol. **7**, 1–205 (1972)
Strobeck, C.: *N* species competition. Ecology **54**, 650–654 (1973)
Thorson, G.: Reproductive and larval ecology of marine bottom invertebrates. Biol. Rev. **25**, 1–45 (1950)
Thorson, G.: Bottom communities. Geol. Soc. Am. Mem. **67** (1), 461–534 (1957)
Thorson, G.: Light as an ecological factor in the dispersal and settlement of larvae of marine bottom invertebrates. Ophelia **1**, 167–208 (1964)
Tikhonov, A.N.: Systems of differential equations with small parameters at the derivatives, (in Russian). Mat. Sb. **31**, 575–586 (1952)
Usher, M.B., Williamson, M.H. (eds.): Ecological Stability. London: Chapman and Hall, 1974
Utida, S.: Interspecific competition between two species of bean weevil. Ecology **34**, 301–307 (1953)

Utida, S.: Population fluctuation, an experimental and theoretical approach. Cold Spring Harbor Symp. Quant. Biol. **22**, 139–151 (1957)
Vandemeer, J. H.: The competitive structure of communities: an experimental approach with protozoa. Ecology **50**, 362–371 (1969)
Veldkamp, H., Jannasch, H. W.: Mixed culture studies with the chemostat. J. Appl. Chem. Biotechnol. **22**, 105–123 (1972)
Verhulst, J. H.: Notice sur la loi que population suit dans son accroissement. Corr. math. phys. **10**, 113–121 (1838)
Volterra, V.: Variazione e fluttuazini del numero d'individui in specie animali conviventi. Mem. Accad. Nazionale Lincei (Ser. VI), **2**, 31–113 (1926)
Wallace, B.: Polymorphism, population size and genetic load. In: Population Biology and Evolution. Lewontin, R. C. (ed.): Syracuse, N. Y.: Syracuse Univ. Press 1968, pp. 87–108
Whittaker, R. H.: A criticism of the plant association and climatic climax concepts. Northwest Sci. **25**, 17–31 (1951)
Williamson, M.: The Analysis of Biological Populations. London: Edward Arnold, 1972
Winogradsky, S.: Microbiologie du Sol: Problèmes et Méthodes. Paris: Masson et Cie., 1949
Wright, S.: Evolution and the Genetics of Populations. The Theory of Gene Frequencies. Chicago: Univ. Chicago Press, 1969, Vol. II

# Subject Index

Vols. 13–21 see next page.